{ An Inordinate Fondness for Beetles }

An Inordinate Fondness *for* BEETLES

Arthur V. Evans *Director·Insect Zoo Natural History Museum of Los Angeles County* Charles L. Bellamy *Senior Curator · Coleoptera · Transvaal Museum Pretoria · South Africa* Photography *by* Lisa Charles Watson *Technical Adviser* Henry Galiano *Illustrations* Patricia Wynne *A* Peter N. Nevraumont Book *A* Henry Holt Reference Book HENRY HOLT AND COMPANY · New York

A Henry Holt Reference Book
Henry Holt and Company, Inc.
Publishers since 1866
115 West 18th Street
New York, New York 10011

Henry Holt® is a registered trademark
of Henry Holt and Company, Inc.

A Peter N. Nevraumont Book

Library of Congress Cataloging-in-Publication Data
Evans, Arthur V.
An inordinate fondness for beetles / Arthur V. Evans,
Charles L. Bellamy; photography by Lisa Charles Watson.—1st ed.
p.   cm.—(A Henry Holt reference book)
Includes bibliographical references (p.) and index.
1. Beetles.   2. Beetles—Pictorial works.   I. Bellamy, C. L.
II. Title.   III. Series.
QL573.E89   1996   96-5245
595.76—dc20   CIP

ISBN 0-8050-3751-9

Henry Holt books are available for special promotions and
premiums. For details contact: Director, Special Markets.

First Edition—1996

Designed by Studio Pepin, Tokyo.

Printed in Hong Kong by Everbest Printing Company
through Four Colour Imports.

All first editions are printed on acid-free paper. ∞

10 9 8 7 6 5 4 3 2 1

Created and Produced by
Nevraumont Publishing Company
New York, New York

Ann J. Perrini, *President*

TABLE OF CONTENTS

PROLOGUE

ASKED what could be inferred about the work of the Creator from a study of His works, the British scientist J. B. S. Haldane is reported to have replied, "an inordinate fondness for beetles." Some people say that Haldane never uttered these words, but no one can argue with the truth they contain. We live in the Age of Beetles. Beetles, or **Coleoptera,** as they are known in scientific circles, inhabit nearly every biological niche, from the narrow fringes of polar ice caps to the broad, unexplored expanse of rain forest canopy. By the most conservative estimate, approximately 350,000 species of beetles have been described since 1758. That's an average of slightly more than four per day. Some of these species, such as the beneficial ladybird beetle and the pestiferous grain beetles, are cosmopolitan. But most species are restricted to a particular continent, mountain range, valley, or soil type.

Using sheer numbers of species as a criterion for success, beetles are the most successful animals on Earth. If single examples of every plant and animal species were placed in a row, every fifth species would be a beetle, every tenth species a weevil (one type of beetle). [PLATE 1] No other group of animals exhibits such a range of size, color, and form. Some beetles are giants; others can crawl through the eye of a needle. Many tropical species are armed with conspicuous horns and claws or resemble jewels and make no attempt to hide themselves; others try to be as inconspicuous as possible, disguising themselves as innocuous materials such as dead leaves or caterpillar feces.

Beetles hold timeless fascination for us. Their peculiar forms and activities have captured the attention of humans for thousands of years and inspired intriguing myths and legends. The ancient Egyptians were fascinated by the size and number of dung-rolling scarabs. The persistent, repetitive behavior of these beetles came to symbolize the invisible forces that move the sun across the sky in a geocentric universe.

For the past 230 million years, beetles have been chewing, digging, and flying through a continually changing landscape. Other creatures share this distinction, but beetles have succeeded at survival as no other group of animals has, as evidenced by the myriad forms and behaviors of beetles that exist today.

Coleopterists, scientists who study beetles, strive to understand the biology of these fascinating creatures. Where do they lay their eggs? How do they develop from wormlike grubs to spectacular adults? What role do they play in their environments? How are different species of beetles related to each other? The study of these specific characteristics of beetles may help answer broader questions: How did they evolve to become the most successful group of animals ever known? What does their success tell us about life on this planet? Can beetles tell us about Earth's past as well as its future? What strategies do beetles use to enhance their survival? Will their success provide a blueprint for human survival?

Most people are not aware of the diversity of beetles and are not impressed by such small creatures—jaded perhaps because many activities of these creatures are considered contrary to human endeavors. Yet we all cross a beetle's path every now and again. This book is for those who have taken a moment to marvel at these fascinating creatures and to wonder where they were going and why. Join us in a trip of discovery as we attempt to fill in some of the many colors on life's palette that are the Coleoptera. We hope you will develop a greater appreciation for beetles and for the scientists who dedicate their lives to studying a particular group of living things. As Albert Einstein said many years ago in concluding an address to the faculty of the California Institute of Technology, "Gentlemen, the deeper I dwell into the sciences of this universe, the more firmly do I believe that one God, or force, or influence has organized all of it for our discovery."

[PLATE 1]

Beetles are the largest group of animals, representing a fifth of all known living organisms and a fourth of all animals. Nearly half of all the beetles are weevils, family Curculionidae, represented here by *Eupholus bennetti* from Papua New Guinea. *Specimen courtesy of Maxilla and Mandible.*

one

〰〰〰

A QUESTION OF NUMBERS

*"Biologists will never be sure that they have found and
named every last species on Earth. But they have a long way
to go before they can even start to wonder."*

•

Nigel Stork and Kevin Gaston, *New Scientist*, 1990

BEETLES, the largest group of insects, representing a fifth of all living organisms and a fourth of all animals, epitomize diversity. Nearly every biological strategy used by terrestrial animal life is represented in this remarkable group of animals that arose during the lower Permian period, about 240 million years ago. Since that time beetles have evolved into nature's single most astounding array of color and form. Their diversity, which eloquently extends beyond the physical, encompassing strategies of behavior, defense, reproduction, and adaptation, has been appreciated since the time of the Pharaohs.

Coleopterists and naturalists have long marveled at how well beetles fit into their environment. Some species are generalists, occupying a variety of habitats; others are restricted to surprisingly narrow environmental parameters. Beetles have penetrated every part of the land that is inhabitable by an insect. Several families, both adults and larvae (the immature forms), have returned to freshwater habitats, occupying fast-moving streams, waterfalls and shallow, murky pools. Only a few species, however, live in or around the sea.

Beetles consume everything—plants, animals, and their remains. Larvae and adults are found in the soil, where they function as tiny recycling machines that return organic materials to the soil, making them available again for use by plants and other animals. [PLATE 2] In turn, beetles themselves are recycled, consumed by many animals, including humans, and occasionally by insectivorous plants.

The teeming multitudes of beetles reflect the degree of their contribution to the health of the biosphere. For most of us, however, contact with beetles is limited to the occasional invasion of our food, homes, and gar-

[PLATE 2]

Beetles have penetrated nearly every terrestrial habitat, consuming plants, animals and their remains. Many species function as minute recycling machines, returning organic materials to the soil for use by other organisms. Top: *Anoplophora medembachi*, Southeast Asia; center left: *Calosoma sycophanta*, France; center: *Batocera laena*, Papua New Guinea; center right: *Eudicella gralli*, Central Africa; bottom left: *Debrobrachus geminatus*; bottom right: *Hybocephalus armatus*, South America. *Specimens courtesy of Maxilla and Mandible.*

dens. Thus we tend to have a limited perspective of these very important creatures. Looking beyond the few pestiferous species, we begin to see the similarities and differences among their hard-shelled throng and instinctively bring order to this chaos in our minds. But before we can begin to explore the unparalleled diversity of beetles, we must look at the system we use to communicate about them.

{ORDER OUT OF CHAOS}

Adam named all of the...animals. Genesis 2:20
The beginning of wisdom is calling things by their right name. Krishtalka, China

SINCE the beginning of recorded history humans have tried to understand their environment better by giving names to things around them. Reportedly, the early Australian Aborigines believed that until objects were named, they did not exist. The ancient Greeks looked for a natural order (*kosmos*) in the diversity of phenomena to identify both animate and inanimate objects. The arrangement of nature by names and categories became the science of **taxonomy** (from the Greek *taxis*, "arrangement," and *nomos*, "law")—the filing system for biology.

In any biological classification scheme, the primary taxonomic unit of importance is the species. A **species** is a group of organisms that have a unique evolutionary history; inhabit a particular geographical range; occupy a biological niche to the exclusion of their nearest relatives; are distinguished by a combination of biochemical, behavioral, and morphological features; and procreate to produce sexually viable offspring. In zoology, species are assigned names, of Latin or Greek derivation, according to the recommendations of the International Code of Zoological Nomenclature adopted by the International Union of Biological Sciences. The code establishes procedures for affixing names to species of animals, including criteria for publishing descriptions of new species and for the availability, validity, and formation of species names. Each new species is identified by a **holotype,** a specimen that serves as the name holder of the species. Investigators who wish to revise the work of previous researchers or to verify the identity of a given species use the holotype as the standard. Most taxonomists prefer to place holotypes in the care of museums or other public institutions so that they are readily available to others for examination.

Once assigned a name, a species is placed within a **classification,** the systematic arrangement of groups of species. With taxonomy providing names, **systematics** adds the dimension of evolutionary relationships to the classification scheme. Systematics began with Aristotle, who in the fourth century B.C. devised groups in which he placed individuals according to a relatively small number of similar characteristics. Aristotle divided animals into two major groups: Vertebrata (the blooded animals) and Invertebrata (the bloodless animals). He then subdivided the Invertebrata into the Malakia (cephalopods and soft-shelled crustaceans), Crustacea (crabs and lobsters), Testacea (most mollusks, echinoderms, ascidians, and other hard-shelled marine animals), and Insecta (insects and arachnids). Aristotle described beetles as insects that have wing cases and thus named them Coleoptera (from the Greek *koleon*, "sheath," and *pteron*, "wing"), which is the scientific name in use today. [PLATE 3]

An increased interest in nature during the Renaissance led to a flurry of

[PLATE 3]

Aristotle described beetles as insects that have wing cases and thus named them Coleoptera, from the Greek *koleon*, "sheath", and *pteron*, "wing". The first pair of flight wings are thickened, protecting the delicate membranous hind wings and the abdomen below, as shown here by a female *Goliathus regius* from Zaire. *Specimen courtesy of Maxilla and Mandible.*

plant and animal descriptions, but without the benefit of a universal systematic framework. The names of many species were actually descriptions in Latin and became so cumbersome that it was almost impossible for naturalists to communicate their findings to one another.

Modern systematics began with the appearance in 1758 of the first edition of *Systema Naturae*, the description of a system of nomenclature developed by the Swedish naturalist Carl von Linné (commonly known as Carolus Linnaeus). Having set out to describe all the organisms of the world, Linnaeus established a system for classifying plants and animals on the basis of comparative anatomy. This system was a major advance in biological thinking. In the Linnaean system, called **binomial nomenclature,** each organism is known by two names, the genus name and the species name. Binomial nomenclature greatly facilitated the storage and retrieval of biological information. Johann Fabricius, an outstanding student of Linnaeus, was the first to classify groups of insects on the basis of their mouthparts. Thereafter, Pierre-André Latreille, Wilhelm Erichson, and Jean Lacordaire continued to transform the Linnaean system, using more and more comparative traits to subdivide groups further and establishing the foundation of our modern beetle classification.

Biological thought developed rapidly in the late eighteenth and early nineteenth centuries, and the work of geologists, particularly James Hutton and Charles Lyell, added the dimension of time. Fossil evidence of strange animals and plants established that entire biotas from earlier times had become extinct. This evidence, coupled with observations on the breeding of domestic animals and the geographic distribution of organisms, independently led Charles Darwin and Alfred Russel Wallace to the theory of transformation of organisms through natural selection. According to the theory of **natural selection,** organisms can slowly change in response to the dynamics of their physical environment. The concept of immutable species, held by the architects of previous systematic schemes, had given way to the idea that species either adapt or become extinct with time.

In his immortal work *On the Origin of Species*, Darwin presented overwhelming evidence that over long periods of time, plants and animals change, ultimately giving rise to new species or ending in extinction. [PLATE 4] Darwin believed that "our classifications will come to be, as far as they can be so made, genealogies." The traditional or classical methods of taxonomy, based solely on physical similarity, were soon transformed into a system based on degree of relationship that used all available attributes, such as morphology, ecology, and distribution. Often, and without foundation, authors proposed new classifications simply because they believed their version to be better than any others. Dissatisfaction with the arbitrariness and lack of methodology in classification led to the development of numerical phenetics and cladistics in the 1950s and 1960s.

Most early attempts at classification were based on physical attributes, or characters, and the degree of physical similarity between organisms. A **character** is any feature used to distinguish one group from another. The grouping of species on the basis of outward appearance is termed **phenetics.** Recognizing the limitation of overall similarity as the sole criterion for classification, several investigators developed a system of using large numbers of characters and recording them numerically to assess overall similarity. This approach later became known as *numerical phenetics*.

[PLATE 4]

The Chilean stag beetle, *Chaisognathus granti,* was included in Darwin's discussion of sexual selection in his 1871 work, *The Descent of Man*. Darwin described the male as "bold and pugnacious". Despite their fearsome appearance, the mandibles of the male are not strong enough to inflict painful or deadly force, but rather are used as forceps for removing rival males from their perches. *Specimen courtesy of Maxilla and Mandible.*

Another advance in systematic theory was a method developed by German entomologist Willi Hennig, who argued that classifications should be based on the branching patterns of shared evolutionary history, which are collectively referred to as **phylogeny.** This method now is generally referred to as **cladistics** because the result of the analysis is a branching diagram, or **cladogram,** which consists of dichotomous branches (that is, branches that split into two directions), or **clades.** The characters that define the phylogeny and branching sequences are categorized as either primitive or advanced. Organisms that share advanced characters are considered to form a natural grouping, sharing a common ancestry. The relationship between organisms that share primitive characters, on the other hand, cannot be determined.

Both the phenetic and cladistic systems have advantages and disadvantages. Each method tries to impose a rigorous examination and careful analysis of all characters, with phenetics attempting to create an information-rich group, while cladistics strives to recreate evolutionary history. A one-sided methodology that does not consider all of an organism's attributes cannot achieve a true classification. Having applied both methods to our own research, we prefer a more traditional perspective. In our view, classification involves a series of steps in which organisms are clustered according to phenetics and their evolutionary branches are determined by cladistics.

Regardless of the methodologies used, biological classifications establish groups that enable biological generalizations, as well as a system that facilitates the filing and retrieval of biological information. Where, then, do people who study beetles conduct their research, and what are some of the basic tools that they use in this endeavor? Coleopterists reside in museums and universities but can just as easily conduct their research in a home laboratory, spending their time studying beetles, both living and deceased, that are pinned, papered, spread, or pickled. Systematists try to establish the evolutionary relationships of an undetermined beetle species and place it within the hierarchical system of the collection they are working on.

All classification categories, or **taxa** (singular *taxon*), above the rank of species are, by definition, subjective groupings that are revised from time to time as new data are considered. Thus their membership expands and contracts. The rank of **genus** (plural *genera*) is immediately above species. Many of the rankings above genus may be identified by their endings; they include tribes (*-ini*), subfamilies (*-inae*), families (*-idae*), and superfamilies (*-oidea*). All of these taxa are grouped into more inclusive groups—classes, phyla, and kingdoms—and each category may be subdivided further. The only taxonomic rank below species that is recognized is subspecies, although this rank has lost favor among many taxonomists.

The most inclusive rank of beetles is the order Coleoptera. [PLATE 5] The Coleoptera are placed in the class Insecta, phylum Arthropoda, and kingdom Animalia. Adult beetles are classified as insects because they have six legs and three major body regions. Like other arthropods (arachnids, millipedes, centipedes, crustaceans, horseshoe crabs), beetles exhibit **bilateral symmetry**—that is, left and right sides that roughly mirror each other—and have segmented bodies and appendages. Beetles are considered animals because they, either as larvae or as adults, ingest complex organic matter, they are capable of moving about their environment, and they exhibit a rapid motor response to stimuli.

[PLATE 5]

The most inclusive rank of beetles is the order Coleoptera, which is further subdivided into evermore exclusive groupings, or taxa. The beetles shown here are all stag beetles of the family Lucanidae. Top: *Chaisognathus granti*, Chile; upper left: *Cyclommatus pulchellus*, Papua New Guinea; center: *Odontolabis lacordairei*, Indonesia; upper right: *Mesotopus tarandus*, Zaire; lower left: *Prosopocoelus bison*, Papua New Guinea; lower right: *Odontolabis femoralis*, Malaysia; bottom: *Dorcus alchides*, Indonesia. *Specimens courtesy of Paul McGray.*

With very few exceptions, members of the order Coleoptera have the leathery front wings known as elytra, and chewing mouthparts. Additional features, such as antennal structure, number of foot segments, connections between body parts, and elytral form, are used to subdivide the order into subordinate taxa. In addition to external characters, beetle systematists rely on the characters of internal morphology, physiology, biochemistry, distribution, biology, and behavior to establish natural subgroups to infer the evolution of the taxon. However, much of the systematic work on beetles is necessarily restricted to external morphological characters, since the biology and fossil record of most beetles remain poorly known or altogether undescribed.

To identify a beetle, a coleopterist compares it to specimens of known identity and checks against published descriptions, illustrations, or photographs, or uses a **dichotomous key.** The key is a concise descriptive device that presents the user with a series of couplets, each consisting of opposing choices based on examination of physical characteristics and distribution. A process of elimination results in identification of the specimen. It is useful to examine many specimens to establish the extremes of sex, size, color, and form of the species under consideration. These studies often reveal species previously unknown to science. Accurate identification is the basis for all biological studies and provides access to all that has been recorded about a particular taxon in the scientific literature. After the identification of a new beetle species has been confirmed or the description of this new species has been published, the new addition to the collection must be curated and catalogued.

Curation of beetles consists primarily of the ordering and care of museum specimens. Specimens are stored in containers to protect them from museum pests. Usually beetle specimens are stored in broad, flat, glass-topped drawers stacked together vertically in cabinets. Specimen collections underpin all endeavors in systematics, providing comparative material for identification, supplying data of historical significance for detecting changes in fauna, and serving as a repository for voucher specimens for species descriptions and ecological surveys.

Systematic collections are like libraries. [PLATE 6] The hierarchical classifications function as call numbers by which we order the specimens, and they help researchers determine how to retrieve additional information in the form of specimens, which are labeled as to their identity, distribution, temporal activity, and habits. Systematic, biological, and zoogeographic information gleaned from these collections are collated into catalogues to document what is known about different groups of taxa. With the huge number of plants and animals already known to science, only listings such as these can help specialists keep track of all the taxa in the group of organisms they are studying.

Our system of classification and collection management is purely a human construct and may reveal more about our own need to view the world around us within a well-defined system than providing a realistic view of biotic relationships. The system imposes a sense of order to the seemingly chaotic universe of beetles and reveals a sense of their relatedness and descent in nature. A true classification—that is, one that accurately portrays the evolution of beetles—is absolutely vital because it generates major

[PLATE 6]

Systematic collections of beetles are like libraries. The hierarchical classification functions as call numbers by which specimens are arranged to assist researchers in retrieving additional information from the scientific literature.
Courtesy Maxilla and Mandible.

ideas about their biology, engendering testable hypotheses about their life cycles, habitats, and behavior.

Like so many other natural resources, beetle collections are threatened with extinction. Shrinking budgets and neglect have destroyed thousands of specimens, and millions more are threatened. The impact of this loss in data is immeasurable. The problem is exacerbated by the fact that coleopterists have poorly communicated to the public how essential beetle systematics is for elucidating biodiversity and ensuring our ability to organize and verify biological knowledge. The neglect of systematic collections has already led to costly mistakes in biological control, such as the introduction of the wrong or ineffectual species to control pests. Analyses of energetics from ecological community studies are virtually useless if organisms under study are not known. Conservationists, in their efforts to develop and enact effective and appropriate legislation to protect species and their habitats, are absolutely dependent on systematics. Collection-based systematic study is essential for developing human perceptions of biodiversity and implementing policies for a sustainable world.

{HOW MANY BEETLES ARE THERE?}

THE question of numbers is not easy to address. Every ten years the United States undertakes a census of all its citizens, and each time there is an outcry over the results. If we cannot count the relatively conspicuous members of our own kind, how can we attach an accurate figure to the total number of living species, many of which are small and secretive? Edward O. Wilson, of Harvard University, offers that we simply do not know, not even to the nearest order of magnitude, and that the number of species could be as high as 10 million or even 100 million! Despite our technology for information retrieval, no database contains the names of all organisms known to science. In the two and a half centuries since Linnaeus set out to describe all the species of plants and animals, only a few of the smaller groups are thought to be nearly completely known. The taxonomy of birds, a group that has been studied intensively by professionals and amateurs alike, appears to be relatively stable. Fewer new species are described each year, suggesting that the catalogue of approximately 9000 species of birds is nearly complete. The list of mammals, encompassing about 4000 species, is also probably near its conclusion.

The group of animals that contains the greatest number of species is the insects. Recent estimates put the total number of insect species known to science at between 1 million and 2 million. Whatever number we pick, we must remember that it is an estimate of the total number of *names*, not species. Some species may have been described more than once; others may encompass several species that await formal description.

With these caveats in mind, how many species of beetles are currently recognized? The number of described beetle species differs in every source that hazards a guess. Modern taxonomic works generally cite the number of beetle species described since the middle of the eighteenth century as between 290,000 and 350,000. Regardless of the number you choose to accept, beetles are undeniably the largest order of animals in the world. [PLATE 7]

Modern scientific descriptions of beetles appeared in literature as early as 1758, with the 654 species described by Linnaeus. Fabricius added

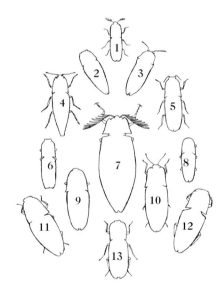

[PLATE 7]

In the 18th century Carl von Linné set out to describe life in its entirety. Yet 242 years later, we still do not know how many species of beetles inhabit our planet. Nevertheless, beetles are acknowledged as one of the most diverse forms of life, as evidenced here by the click beetles, family Elateridae.
1. *Abiphis insignis*, Madagascar; 2. *Campsosternus* sp., Malaysia; 3. *Campsosternus gemma*; 4. *Semiotus* sp., Colombia; Taiwan; 5. *Chalcolipidius* sp., Brazil; 6. *Chalcolipidius smaragdinus*, USA; 7. *Oxynopterus mucronatus*, Malaysia; 8. *Chalcolipidius webbi*, USA; 9. *Chalcolipidius* sp., Columbia; 10. undetermined species, Papua New Guinea; 11. *Chalcolipidius rugas*, Mexico; 12. *Lycoreus alluaudi*, Madagascar; 13. *Alaus oculatus*, USA. *Specimens courtesy of Paul McGray.*

another 4112 species between 1775 and 1801. Thousands of descriptions of beetles began to appear at the beginning of the nineteenth century, compelling German coleopterists Max Gemminger and Edgar von Harold between 1868 and 1876 to compile a catalogue of beetles that contained nearly 77,000 species. The successor to this catalogue, which was published by W. Junk and Sigmund Schenkling between 1910 and 1940, with occasional supplements, listed nearly 221,500 species. The Junk and Schenkling tomes represent the last attempt at a complete catalogue of all the world's beetles.

With hundreds of thousands of species known to science, it has become convenient to discuss beetles in terms of the families they have been assigned to. Appendix 1 gives the most recent accounting of beetle families. Approximately two-thirds of the known beetle species reside in just eight families: Buprestidae (jewel beetles), Carabidae (ground beetles), Cerambycidae (long-horned beetles), Chrysomelidae (leaf beetles), Curculionidae (weevils), Scarabaeidae (dung beetles), Staphylinidae (rove beetles), and Tenebrionidae (darkling beetles). [PLATE 8] Of these, the largest family is the curculionids, comprising more than 50,000 named species.

To resolve the discrepancy between our catalogues and the real world, coleopterists must reexamine collections and continue to explore remote regions and habitats in search of new species. In almost every insect collection, whether at a small college or at a large public institution, it is not surprising to discover, within drawers of unidentified specimens, species previously unknown to science. As in archaeology, many entomological (insect) treasures have been unearthed and await the attention of trained specialists to interpret and document their value.

The Musée National d'Histoire Naturelle in Paris reportedly holds the largest insect collection in the world—perhaps as many as 100 million specimens representing 400,000 species. Tucked away in this incredible collection is an accumulated wealth of undescribed coroebine buprestids, a group of jewel beetles. These specimens were first collected by a group of ground-nesting wasps as prey in different habitats throughout Madagascar. By noting the habits of these wasps and digging out their burrows, French and Madagascan collectors unearthed a collection of beetles amounting to tens of thousands of specimens estimated to represent more than 500 undescribed species and more than twenty new genera. A significant proportion of these buprestids are possibly extinct as a result of the destruction of up to 80 percent of the original forest cover.

The inaccessible tropical rain forest canopy is poorly explored and has been dubbed by Terry Erwin, a coleopterist from the Smithsonian Institution, the "last biological frontier." The use of aerial insecticides that contain toxins that break down quickly in the environment enable the collection of nearly 100 percent of the arthropod fauna within the treated area. After being collected, these specimens are processed and shipped to research institutions for study by specialists.

From this species-rich habitat came the first stunning extrapolation of the extent of beetle diversity. In 1982, using data collected in Panama from nineteen individuals of one species of forest tree (*Luehea seemanii*, family Tiliaceae) in a seasonal wet forest, Erwin estimated that there are more than 8 million species of beetles in the tropics alone. His extrapolation was based on simple mathematical assumptions. First, he assumed that, on the basis of

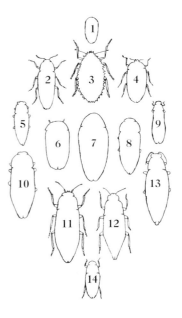

[PLATE 8]

Approximately two-thirds of the known beetle species reside in just eight families, including the Buprestidae. The buprestids, commonly known as metallic wood boring or jewel beetles, are popular with collectors because of their metallic, streamlined bodies. 1. *Stigmodera gratiosa*, Australia; 2. *Cyria imperialis*, Australia; 3. *Julodis viridipes*, South Africa; 4. *Stigmodera roei*, Australia; 5. *Chalcophora japonica oshimana*, Japan; 6. *Stigmodera macularia*, Australia; 7. *Sternocera aurosignata*, southeast Asia; 8. *Themognatha chalcodera*, Australia; 9. *Demochroa gratiosa*, Malaysia; 10. *Themognatha ducalis*, Australia; 11. *Lampropepla rothschildi*, Madagascar; 12. *Cyphogastra javanica*, Indonesia; 13. *Chrysochroa fulgidissima*, Taiwan; 14. *Lampetis* sp., southeast Asia. *Specimens courtesy by Gerald Larsson and Paul McGray.*

feeding strategies of the 1200 different species of beetles collected, 163 were found only on this one kind of tree. Second, Erwin assumed that this number of host-specific species was average for all tropical tree species and thus multiplied the average number of host-specific beetles by 50,000, the approximate number of tropical tree species. The result, representing the total number of beetle species, was 8,150,000. Since beetles represented about 40 percent of the tree's total insect diversity, Erwin extrapolated that the total number of insect species in tropical rain forest canopies was about 20 million!

Erwin's paper explaining this extrapolation spawned a worldwide cottage industry of studies to support, modify, or refute his findings. Estimates for the total number of insect species now vary from 1.87 million to 80 million! Researchers continue to refine arthropod sampling techniques and the methods of analysis for determining the relationship of each species to the host plant to estimate more accurately the number of species supported by the tree. The goal is to be able to make more reliable extrapolations regarding trees within similar forest habitats.

There are three good reasons to pursue the quest of determining the total number of species of beetles. First, simply knowing how many species there are should be an end in itself. In the space program, for example, billions of dollars have been spent just to see what is out there. Why can't taxonomists lobby the government to support the exploration of our own planet for the same reason? Second, understanding the number of species involved can help us assess the job that taxonomists have been doing since Linnaeus threw down the gauntlet to describe life in its entirety. Finding out where we are in this process and attempting to achieve a balance among groups and regions requiring further study will help us chart the course of biological exploration well into the new millennium. Finally, the continuing devastation of huge tracts of tropical forests and other poorly known habitats by chain saws, bulldozers, and bonfires lends urgency to the question, How many species of beetles are we losing every year? An answer to this question requires knowing the total number of living species.

Our inability to come up with a reliable estimate of the number of beetle species in the world highlights the fact that we know very little about this important group of animals. No matter how many beetles await a systematic accounting, time and funds are in short supply, increasing the likelihood that species will become extinct in nature before we have the opportunity to name them. However, the discovery and subsequent description of new species do not constitute knowledge; they are simply the jumping-off point for further investigation. The mere naming of a species fails to reveal anything about its food preferences, developmental stages, defense mechanisms, interactions with other organisms, or other unique or mundane characteristics that are at the core of its own special nature. The disparity in the estimates of the numbers of beetles brings to our attention that they and other invertebrates have long been neglected, a puzzling fact, since we possess enough data to know that their fate is central to the biodiversity crisis.

{ THE SECRETS OF SUCCESS }

TWO key factors have contributed to the overall success of beetles: their body forms and their reciprocal interactions with plants. The

[PLATE 9]

The tough and sometimes brightly colored exoskeleton of beetles, such as *Jumnos ruckeri* from Thailand, serve to protect them from dessication and abrasion. *Specimen courtesy of Maxilla and Mandible.*

[PLATE 10]

A flower chafer, *Euphoria sepulchralis*, from Arizona, feeding
on pollen. *Photograph by Charles Bellamy.*

A large jewel beetle from Malawi, *Sternocera orissa variabilis*,
feeding on the flowers of *Acacia*. *Photograph by
Charles Bellamy.*

body adaptations that beetles exhibit reveal a remarkable collective ability to reconcile their needs to a continually changing environment. These highly specialized bodies have permitted beetles to occupy a staggering array of habitats with relative ease. Small and compact, beetles are well equipped to hide, search for food, and lay eggs by burrowing in the soil or slipping into the narrow cavities among stones and rotten logs.

The tough external skeleton, called an **exoskeleton,** protects the beetle by eliminating exposed membranes and thereby reducing desiccation. Thickened wing covers called **elytra** (singular *elytron*) protect beetles from abrasion as they move about through soil, debris, and decomposing plant materials and help shield them from the ravages of predators and parasites. [PLATE 9] These elytra usually cover most of the body, protecting its delicate contents. A specialized cavity beneath the elytra enables some beetles to exploit desert habitats, insulating their bodies from the heat and minimizing water loss through respiration. This cavity can also store oxygen, allowing some species to take advantage of aquatic habitats. Beetles are not particularly fast or agile fliers; they have been said to resemble airborne trucks in low gear. Nevertheless, taking to the air enables beetles to avoid predation, locate mates, find food, and colonize new and unexploited habitats.

The crucible of interactions between plants and beetles has contributed to the overall success of beetles. [PLATES 10 AND 11] Beginning with their preference for fungi some 240 million years before present, beetles were predisposed to exploit the new food source that appeared on the scene about 125 million years before present, the flowering plants. Plants are not passive partners in this relationship; they respond to herbivory (plant eating) with both physical and chemical feeding deterrents. Not easily put off, however, beetles adapt strategies to circumvent the defense mechanisms of the plant. Plants, in turn, come up with new defenses. This ancient dance between plants and beetles, which continues today, has led beetles down an unprecedented path of evolutionary possibilities and opportunities, resulting in their emergence as one of nature's most successful designs.

two

THE BEETLE BLUEPRINT

The beetles are at once absolutely typical of,
and unique among, the Insects.

•

Roy Crowson, *The Biology of the Coleoptera*

{THEME AND VARIATIONS}

COLEOPTERISTS have described about 350,000 variations on what we easily call a beetle. The bewildering variety of beetle body form is a testament to their unmitigated success on Earth and reveals strategies, wrought from millions of years of success and failure, of populations attempting to come to terms with ever-present biotic and abiotic environmental challenges, for beetles are evolving not *in* the world, but *with* it.

The armored bodies of adult beetles make them conspicuous. They are easily recognized, even by the casual observer. Yet the diversity in beetle form is mind-boggling, an array that at first glance seems alien to our relatively narrow and decidedly vertebrate perception of life. The smallest beetles, known as feather-winged beetles, or ptiliids, are 0.035 millimeters long [ILLUSTRATION A] and could comfortably undergo their entire life cycle within the head capsule of one of the largest cerambycid beetles, *Titanus giganteus*, which can grow as long as 200 millimeters. [PLATES 12 AND 13] The colors of beetles exceed the palette of the most avant garde of painters.

Despite their diversity, beetle bodies share common characteristics that clearly identify them as beetles. These physical attributes serve as a template for our scheme of classification, which strives to reveal the evolutionary relationships of beetles.

{DRESSED FOR SUCCESS}

BEETLES have systematically been placed among the arthropods in part because they are contained within a highly modified external cylinder, the **exoskeleton.** The exoskeleton is one of the primary organ systems, functioning as both protection (like skin) and structure (like a skeleton). Each of its many layers consists primarily of chitin and protein.

actual size

[ILLUSTRATION A]

One of the smallest beetles in the world, *Acrotrichis* sp. from North America, is a diminutive 0.05 mm in length. The smallest known beetle in the world is *Nanosella fungi*, found in eastern United States, measuring in at a whopping 0.035 mm!

[PLATE 12 *opposite*]

Titanus giganteus is one of the largest beetles in the world, measuring from 12.0–20.0 cm in length. Less than 20 specimens were known in collections before 1957. Since then, hundreds of specimens have been collected in northern Brazil and French Guiana, yet still nothing is known of their biology.
Specimen courtesy of Maxilla and Mandible.

[PLATE 13]

Two of the world's largest beetles, *Titanus giganteus* and *Callipogon armillatus*, can easily fill the palm of an adult human hand. Both of these specimens were collected in French Guiana. *Photograph by Rosser Garrison.*

[PLATE 14]

The outer surface of the exoskeleton may be covered with
spines, or hairlike structures, or setae, as seen here on this
buprestid from Arizona, *Acmaeodera gibbula*. Setae may
function in a sensory capacity, transmitting environmental
information to the nervous system, or they may to protect
against predators, abrasion, and dessication.
Photograph by Rosser Garrison.

Chitin, first identified in 1823, is fibrous and makes the exoskeleton tough, yet flexible—not unlike fiberglass. The outer layers of the exoskeleton are collectively called the **cuticle.**

In adult and larval beetles the outer surface of the exoskeleton may be covered with spines, or **setae** or of various functions and configurations, or coated with waxy secretions. [PLATE 14] These adornments may function in a sensory capacity, transmitting tactile or environmental information to the nervous system, or they may help protect against predators, abrasion, and desiccation. The texture of the exoskeletal surface reveals incredibly diverse microsculpturing, which, when magnified many times, resembles a surreal extraterrestrial landscape. Complex infoldings of the exoskeleton create an internal framework to which powerful muscles attach and from which they exert leverage.

The thickness and durability of the exoskeleton affords some protection from predators and other potentially life-threatening predicaments. The ant-acacia scarab *Pelidnota punctulata*, for example, is a conspicuous ruteline scarab that lives in the lowland forests of Costa Rica. Nocturnal, these beetles feed primarily at night on the tender shoots of ant acacias (*Acacia cornigera, Acacia collinsii*), and they cling to the plant during the day. Ant acacias are so named because of their impressive defense system: They house a colony of biting and stinging acacia ants of the genus *Pseudomyrmex*, which occupy the swollen, hollow thorns of the tree. The acacia provides all the nutritional requirements of the ants. In return, the ants attack any herbivore (plant eater) that attempts to make a meal out of their arboreal home, including *Pelidnota punctulata*. Although nearly always under siege by the ants, this scarab beetle never tries to flee. Given that other species of similarly armored *Pelidnota* immediately drop to the ground when attacked by these ants, *P. punctulata* must be well protected against ant attacks, and the behavior of this species must also have been modified.

The ironclad beetle, *Asbolus verrucosus*, a tenebrionid that inhabits the drier portions of some North American deserts, also possesses a tough exoskeleton. Its roughened cuticle, which resembles the impenetrable metallic exterior of the infamous metal warships of the American Civil War, help the ironclad beetle live up to its name. The primary purpose of this beetle's hardened cuticle is to protect against loss of the most precious of all desert commodities, water. A bluish gray waxy substance secreted by glands that have ducts at the tip of knoblike projections on the elytra coats the ironclad beetle's exoskeleton. The thickened cuticle and waxy film combine to form something like a deep-sea diving suit in reverse, designed to retain moisture rather than keep moisture out. The light color of the wax also reflects more sunlight than the surrounding darker cuticle, helping to keep the body cool.

To facilitate flexibility, the exoskeleton consists of a series of segments that are subdivided into individual plates, called **sclerites.** Sclerites are separated from one another by a thin, flexible groove in the exoskeleton or by a membrane of pure chitin. The sclerites themselves contain chitin and are impregnated with a protein called **sclerotin.** The cross linkage between the molecular chains of chitin and sclerotin give the thin, light sclerites amazing strength.

The body of the beetle consists of three main segments: **head, thorax,** and **abdomen.** [PLATE 15] Viewed from above, adult beetles show three

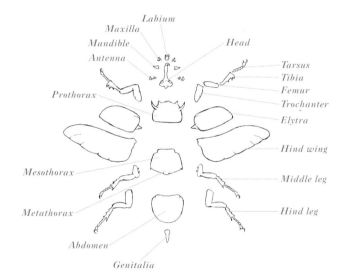

prominent features: head, prothorax, and elytra (the thick wings). The head bears the eyes, antennae, and chewing mouthparts, which are present on almost all beetles. Although, as in other insects, the **prothorax** is connected to the remaining two thoracic segments, the **mesothorax** and the **metathorax**, in beetles these last two thoracic segments are covered by the elytra. The fundamental thoracic and abdominal segments consist of four sclerites arranged in a ring, the dorsal tergum (**tergites**), two lateral pleura (**pleurites**), and the ventral sternum (**sternites**). [PLATE 16] Segmented appendages manifest themselves as mouthparts, legs, wings, copulatory and egg-laying organs. All beetles have this basic body plan; only the details vary.

The basic adult beetle body plan resembles the form of some armored vertebrates—armadillos, turtles, and tortoises. No beetles show this similarity more clearly than the cassidine beetles of the family Chrysomelidae, aptly named the tortoise beetles. Many significant examples, however, deviate from the smooth, humpbacked forms that, for most people, typify beetles. For example, many aquatic beetles, most of which belong to the families Gyrinidae, Hydrophilidae, and Dytiscidae, sport hydrodynamic bodies that are compressed top to bottom and keeled laterally and occasionally ventrally, as evidenced by the presence of a ventral spine in some hydrophilids. The projectile-like bodies of buprestid beetles and the stout cylindrical body plan of many members of the wood-boring Cerambycidae, with their short powerful mandibles (jaws) mounted on the front of the head, eliminate much of the time and energy of excavating large galleries in solid wood, which would be required to accommodate less streamlined bodies.

Mormolyce phyllodes, the fiddle beetle, is strangely compressed top to bottom. [PLATE 17] The flattened torso is a result of this beetle's adaptation to hunting for insects in the narrow space between the bark and wood of decaying logs. The predaceous South African scarab *Placodidius* is paper thin when viewed laterally and is known to inhabit the narrow passages of termite galleries. Dung scarabs, which, to avoid competition from flies and other beetles, must quickly hide the feces that are their food, are modified to resemble earth-moving equipment in miniature, complete with shovel-like forelegs, scrapers attached to the head, and thoracic scoops to loosen and remove soil.

Other morphological extremes in beetles include the antlike shape of anthicid beetles, whose form and rapid movements may confuse predators into thinking that they are their more pugnacious models. Some staphylinids and aphodiine scarabs are part of what has been called the "social insect's bestiary" and have evolved bizarre abdominal modifications that give them a close physical resemblance to that of their ant and termite hosts. This strategy is not intended to repel their enemies, but rather facilitates their integration among their hosts.

{ SETTING THE STAGE }

BEETLES have four main life stages—egg, larva, pupa, and adult—each with distinctive characteristics. Beetle eggs are usually soft and smooth. The shell, or **chorion,** is permeable to water and, in the case of smaller eggs, oxygen. Gases pass through the chorion via **aeropyles.** Another aperture in the chorion is the **micropyle,** a series of pores through which

38

[PLATE 17, *opposite*]

The Javan fiddle beetle, *Mormolyce phyllodes*, is strangely
compressed, top to bottom. This adaptation allows the fiddle
beetle to move easily underneath the bark of fallen
trees in search of insect prey. *Specimen courtesy of Paul McGray.*

[PLATE 18]

Beetles function in the environment as two distinct animals—
the larva and adult. The word larvae, derived from the Latin
for "ghost" or "mask", probably was applied to immature
insects because of their secretive habits. The larvae of the
melolonthine scarab, *Polyphylla erratica*, leads a subterranean
existence along the Armagosa River in the California and
Nevada desert, feeding on the roots of salt grass.
Photograph by Arthur Evans.

sperm (the male gametes) that have been stored in the mature female reproductive system enter as the female lays the eggs.

The word **larva,** derived from the Latin for "ghost" or "mask," probably was applied to immature insects because of their secretive habits. [PLATE 18] Beetle larvae usually look like grubs or worms, but some predatory forms may be flattened and have distinct legs. The larvae have well-developed, sclerotized head capsules fitted with chewing mouthparts. The antennae vary but usually have two to four segments. The head bears an inverted Y-shaped groove, called the **ecdysial suture,** which is a line of weakness that functions as an escape hatch, during **molting,** allowing the larva to withdraw from the exoskeleton it has outgrown. Most beetle larvae have from one to six simple eyes on each side of the head; a few have none. The mouthparts may be modified for crushing, grinding, tearing, or, rarely, sucking.

The three thoracic segments are very similar to one another, but the prothorax may have a more heavily sclerotized dorsal plate than the mesothorax or metathorax. The legs, when present, may have six (suborder Adephaga), five (suborder Polyphaga), or fewer segments. The abdomen usually has ten segments but may have nine or eight, and it is usually completely membranous. Although the abdomen lacks legs, its segments may be endowed with ambulatory warts, called **ampullae.** These sclerotized blisters on the abdominal surface help the larva to gain leverage while moving. Prolegs may appear on the first two abdominal segments, and the anus is sometimes surrounded by a fleshy lobe that forms an anal foot or pad. The ninth abdominal segment may have a pair of fixed or jointed appendages toward the back, known as **urogomphi.**

The word **pupa** is derived from the Latin for "little girl" or "little doll." The pupae of beetles look somewhat like dolls and are generally **exarate**— that is, the legs and wings are not cemented to the body. The pupal abdomen may have functional muscles, allowing for some movement. **Gin-traps** are sclerotized teeth that form in some species on the opposite surfaces of some of the abdominal segments and that can be made to close together sharply by sudden contraction of the dorsal longitudinal muscles. These devices are thought to help protect the grooves between segments of the abdomen, which are often targets for small predators and parasitic mites.

{MEETING THE WORLD HEAD ON}

THE beetle head is generally a hardened capsule attached to the thorax by a flexible, membranous neck. The head capsule is usually the first part of the beetle to come into contact with the environment. It houses delicate and primary sensory structures such as the eyes, antennae, and brain, as well as the opening to the pathway for obtaining nutrients, the mouth. Although it lacks any outward appearance of segmentation, the head has a groove, called the **frontoclypeal suture,** that marks an infolding of exoskeleton that forms part of a group of internal struts collectively known as the **tentorium.** The tentorium functions as an internal brace and as a point of attachment for the incredibly powerful muscles of the mandibles.

The upper anterior portion of the head, called the **clypeus,** supports the **labrum,** or upper lip. In weevils and some other beetles, the labrum is absent. The labrum covers the rest of the beetle's mouthparts: two pairs of jaws, called the **mandibles** and the **maxillae;** and the lower lip, or **labium.** The mandibles are usually conspicuous, curved, and sometimes toothed along the interior edge and may be monstrously developed in the males of some lucanids and cerambycids, suggesting the antlers of a stag. [PLATES 19 AND 20] The mandibles are modified to cut, grind, or strain foodstuffs and may occasionally serve as a means of defense. Male tiger beetles use their mandibles to restrain the female during copulation. The maxillae and labium may have delicate fingerlike structures, **palpi,** that facilitate the intake of food, much in the way that we use our fingers to manipulate food.

The mouthparts of beetles are usually protected underneath by a plate known as the **gula,** a structure that is characteristic of the coleopterous head. Although the mouthparts of chrysomeloids and curculionoids are directed downward (**hypognathous**), the mouthparts of most beetles are directed forward, parallel to the axis of the body (**prognathous**).

The heads of beetles are modified in various ways. Nowhere is such modification more pronounced than in the curculionoids—the weevils and their relatives. The mouthparts of these beetles are mounted at the very tip of an extended rostrum (snoutlike protuberance), often with the antennae inserted not far behind. [PLATES 21 AND 22] Other families of beetles exhibit similar but somewhat less spectacular modifications, with the antennal insertions nearer to the eyes. These adaptations are often associated with beetles that feed on nectar within flowers.

FINDING THEIR WAY ON the head of the beetle is a pair of **antennae,** each of which is attached at a point between the base of the mandible and the compound eyes. [PLATE 23] The edge of the compound eyes of most cerambycid beetles is kidney-shaped, modified to accommodate the insertion of the antennae. This adaptation may reflect the importance of this particular part of the beetle's anatomy as a sensory site for touch and smell. The antennae are equipped with incredibly sophisticated sensors that detect food, locate egg-laying sites, detect vibrations, and assess environmental factors such as temperature and humidity. To maintain their sensitivity, some carabids and staphylinids possess specialized structures on their legs that are used to clean the antennae regularly.

The antennae are generally shorter than the body, although some cerambycids and brentids have antennae that may reach three times their body length. In some families, the antennae of females differ considerably from those of males. [PLATE 24] Males may have very elaborate antennal structures to increase their powers of smell. [PLATE 25] Such adaptation is especially evident in species that use **pheromones,** external hormones that communicate messages to other members of the species.

Although the antennae appear to be segmented, generally they lack internal musculature, instead having ringlike structures called **annuli.** The basic number of annuli for beetles is 11, but reductions to 10, 9, and 8 annuli are common, while still other taxa may have 7 or fewer segments, or more that 30, as in the males of some Rhipiceridae. Each antenna consists of a basal **scape** that is connected to a smaller structure called the **pedicel.** The scape is inserted in the **antennifer,** a ball-and-socket–type extension that enables easy rotation of the antenna in all directions. The antennae move

[PLATES 19 AND 20]

The mandibles of most beetles are conspicuous and
occasionally, as in these cerambycids, somewhat monstrous.
As its specific name suggests, the Central American *Callipogon
barbatus* has a pair of well-developed mandibles covered with
thick pile, suggesting a beard. *Macrodontia cervicornis*, Brazil,
bears mandibles which are conspicuously toothed.
Specimens courtesy of Paul McGray.

[PLATES 21 AND 22]

One of the most striking modifications of the beetle head is
seen in the curculionoid beetles, weevils and their relatives.
Above: an unidentified weevil from French Guiana;
Right: a brenthid, *Belopherus maculatus*, Puerto Rico.
Photographs by Rosser Garrison.

[PLATE 23]

On the head of a beetle is a pair of antennae, each of which is
attached at a point between the base of the mandibles and the
compound eyes, as seen here in this unidentified prionine
cerambycid from peninsular Malyasia.
Photograph by Charles Bellamy.

The antennae of most beetles are generally shorter than the body, although some cerambycids and brentids have antennae that may reach three times their body length. In cerambycids, such as *Pseudomyagrus waterhousei* from Malaysia, the antennae of females are considerably shorter than those of the males. *Specimens courtesty of Maxilla and Mandible.*

[PLATE 25]

The antennae of the male elaterid, *Hemiops nigripes*, and cerambycid, *Diastocera wallichi*, are highly modified structures designed to increase their sense of smell. These adaptations are often indicative of the species' reliance upon pheromones, airborne messengers used to locate mates buried in the soil or hidden among tangled vegetation. *Specimens courtesy of Maxilla and Mondible.*

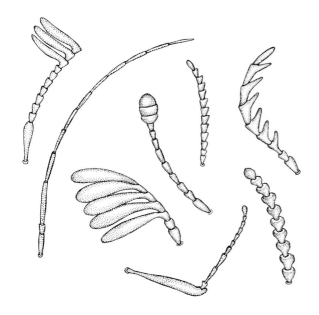

[ILLUSTRATION B]

The antennae of beetles are packed with incredibly sensitive chemo-receptors used to detect food, assess environmental conditions and locate mates. Coleopterists may also use antennal structures to place beetles within the hierachical classification. From left to right: lamellate, filiform, geniculate, flabellate, capitate serrate, captiate, moniliform, serrate.

[ILLUSTRATION C]

The flattened or lamellate segments of the antennae of scarabs, such as the nocturnal *Polyphylla fullo*, may be expanded like a fan or folded tightly into a compact club. This feature distinguishes scarabs from all other beetles.

via muscles that originate on the tentorium and are inserted in the pedicel. The scape and pedicel also contain the complex **Johnston's organ**, which generally consists of organized groups of sensory cells called **scolopidia**. The Johnston's organ is weakly developed in most adult beetles, but whirligigs (family Gyrinidae) are an exception. In whirligigs this organ is highly developed and probably functions as part of a system to detect and locate the source of vibrations on the water surface emanating from the thrashings of potential predators and prey, or the energy patterns emitted by other whirligigs.

The annuli, which lack any internal musculature and are joined to one another by extremely flexible membranes, are collectively known as the **flagellum**. The flagella exhibit an amazing array of modifications that are characteristic of families of beetles and have been variously described as filiform (long and slender), moniliform (short and beadlike), serrate (saw-toothed), pectinate (comblike), flabellate (extreme comblike or featherlike), clavate or capitate (distinctly clubbed), lamellate (terminal annuli flattened or platelike), and geniculate (jointed between the scape and annuli). [ILLUSTRATION B]

The protective adaptations of species that occupy hostile or abrasive environments emphasize the importance of antennae to beetles. Many species that live in symbiotic relationships with ants or termites have grooves into which the antennae and other appendages may be partially or completely retracted, thereby reducing the chances that the beetle will be attacked, ejected, or killed by its hosts. The delicate platelike lamellae of most scarabaeoid beetles may be expanded or retracted into a club, but in some burrowing geotrupids the distal lamellae may be nearly completely enveloped by the proximal cup-shaped lamella. [ILLUSTRATION C] Nitidulid sap beetles can expand the annuli of their capitate antennae to expose sensor-laden inner surfaces.

Antennae may also function in other capacities. Water scavenger beetles of the family Hydrophilidae trap a layer of air over the ventral surface of their body by first breaking the surface tension of water with an antenna. In larval *Hydrophilus*, the spiny antennae actually assist the mandibles in the chewing of captured prey. The neotropical cerambycid *Hammaticherus* uses its elongate antennae as weapons of defense. Both males and females of the species *Hammaticherus batus* have several antennal annuli armed with long recurved spines. When threatened, the beetle whips its antennae about, backward and downward, bringing the spines into contact with the attacker. A twist of the head will embed these spines into the flesh of a would-be collector, causing pain and minor loss of blood.

The antennae of paussine carabid beetles are characterized by a reduced number of annuli. The few that are present are fused into a flattened terminal club, which may be relatively massive in some species. This highly modified club is filled with gland cells that are believed to secrete substances that increase the tolerance of paussines by their ant hosts.

WINDOWS TO
THE WORLD

BEETLES have a pair of hemispherical **compound eyes** that in most families are broadly separated from one another. The edges of the eyes may be unbroken, notched, or kidney-shaped, or they may be partially or completely divided by a ridge of cuticle known as the **canthus.** Many flightless beetles have reduced compound eyes, and some species that live in caves or in the litter of the forest floor, adapted to living in total darkness, lack them altogether. Simple eyes, called **ocelli,** are rare in adult beetles. Most dermestids and one staphylinid have a single ocellus in the middle of the head, and some genera of leiodids, omaliine staphylinids, and derodontids have two.

Each compound eye is made up of dozens or hundreds of individual facets. Each facet is part of a complete visual unit, or **ommatidium,** which consists of several elements. The facets of the eye are its corneal lens and are made up of colorless cuticle and structurally modified according to the beetle's mode of life. Relatively small, flat facets are sufficient to illuminate the world of most diurnal beetles; larger, more convex facets are common in nocturnal species. The **crystalline cone** is a generally blunt cone that is distinct from the corneal lens. It is surrounded by four cone cells, which are clear and are in turn surrounded by two or more sheathing pigment cells. The migrations of pigment within these cells act as an iris, regulating the amount of light coming in and adapting the eye to light or dark.

Beetles live in a world of light that humans can perceive only indirectly. Beetles can see ultraviolet or infrared light, but exactly how they view the world is open for debate, since the resultant image depends on how the brain processes the nerve impulses generated by light coming in from the ommatidia. Most beetles probably view the world as a single image made up of many pieces, as opposed to a mosaic of multiple images.

One of the most extreme developments of beetle visual equipment is demonstrated by the gyrinids, or whirligigs. Whirligigs literally live on the edge, gyrating about the thin surface film of ponds and slow-moving streams in search of food and mates. These animals benefit greatly from an enhanced ability to view their world above and below with equal clarity, an ability afforded them by the possession of four eyes. [ILLUSTRATION D] Each compound eye is divided into two parts by a strip of cuticle. The exposed upper portion is designed for viewing the scene through relatively dry air, while the submerged lower portion is arranged to gather an image through the thicker, wetter medium below.

[ILLUSTRATION D]

Whirligig beetles, such as members of the genus *Dineutes*, are supremely adapted to living on the edge of their aquatic environment, as demonstrated by their unusual eyes. Each compound eye is divided into two parts. The exposed upper portion of the eyes are designed to collect images above the surface, while the submerged portion is adapted for seeing in the thicker, wetter medium below.

BIGGER IS NOT
NECESSARILY BETTER
THE most extraordinary and controversial modification of the beetle head is that of horns. [PLATE 26] In both mammals and beetles, horns have evolved independently many times. Horned beetles are adorned with extrusions from the head and/or thorax that may suggest antlers or tusks. [PLATE 27] The horns of some beetles are enormous, but even within species there may be considerable variation. Horns may be knoblike structures that jut out from the dorsal surface of the prothorax or form a crown on the head that jabs forward toward unknown enemies or sweeps forcefully and symmetrically back over the body. Horns may arise from other parts of the head, mandibles, forelegs, and elytra. In some species, structures on the head and thorax may work in precise concert, forming a vise of sorts; in others, the relationship between similar structures seems more haphazard to the casual observer. Although horns generally adorn male beetles only, the females of the African dung scarab *Heteronitis castlenaui* proudly display a broad hoelike adornment while the males lack any such structure.

Horned beetles are the subject of much debate and speculation, but

[PLATE 26]

One of the most extraordinary modifications of the beetle head is that of horns. Left to right, top to bottom: *Dicronorhina derbyana*, South Africa; *Eudicella gralli*, Zaire; *Eudicella ducalis*, Burundi; and *Eudicella immaculata*, Kenya. *Specimens courtesy of Paul McGray.*

[PLATE 27 *above*]

The horned cetoniine scarab, *Anisorrhina algoensis*, from South Africa. *Photograph by Charles Bellamy.*

[PLATE 28]

Charles Darwin suggested that if the atlas beetle, such as this *Chalcosoma caucasus*, were magnified to the size of a dog or horse "with its polished bronze coat of mail and its vast complex of horns...it would be one of the most imposing animals in the world." *Specimen courtesy of Maxilla and Mandible.*

seldom are they the focus of serious investigation. In mammals, horns, antlers, or tusks serve as weapons to be used in combat usually against members of the same species, in which the participants attempt to stab or gore one another. The heavily chitinized exoskeleton of most beetles would seem to preclude this kind of violent application of horns. Nevertheless, some biologists have suggested that enlarged structures of the head and thorax function as weapons, enhancing the reproductive capability of the bearer by improving their chances in battles against other males. Charles Darwin suggested that if the male *Chalcosoma* were magnified to the size of a dog or a horse, "with its polished bronze coat of mail and its vast complex of horns . . . it would be one of the most imposing animals in the world." [PLATE 28] Darwin apparently subscribed to the idea that bigger is better, and he suggested that female beetles prefer males with bigger horns, but such a preference has not been documented.

The smaller males of the bearded weevil, *Rhinostomus barbirostris*, avoid the aggressive behavior of their larger brothers and cousins by capitalizing on their diminutive size. They have been observed to dart past larger, more distracted males who may be jousting one another, to inseminate the female who inspired the combative behavior of the larger males, thus ensuring their contribution to the gene pool.

Gilbert Arrow, a renowned coleopterist at the British Museum of Natural History for more than fifty years, compiled an exhaustive work, published posthumously, on horned beetles. Arrow concluded that for the most part horns have no function and that their fantastic development would preclude their effectiveness in battle. He proposed that the evolution of horns was incidentally linked to increased body size of the individual beetle.

Today, evolutionary biologists attempt to view the development of any natural feature within the constraints of natural selection. Through careful observation they explore the relationship between form and function in these structures to explain why a particular configuration might develop. Several factors play a role in the development of horns in beetles. Variation of horn and mandible development in beetles may or may not be associated with overall body size. Nutrition and other environmental parameters also influence the development of horns. For most species of horned beetles though, the significance of the range of horn size within a species remains obscure. Most observations indicate that the vast majority of horned beetles use their protuberances to rid themselves of competitors for females and for food.

Horned beetles are poorly studied because of their cryptic or nocturnal behavior and/or because they tend to inhabit remote parts of the globe. But the forked fungus beetle, *Bolitotherus cornutus*, less than an inch long, has proven to be an ideal subject for horned beetle study. [ILLUSTRATION E] Forked fungus beetles inhabit deciduous hardwood forests in the northeastern United States and may live up to five years. Horn size ranges dramatically in this species: Major males bear horns that are half the length of their abdomen, while the horns of minor males may be practically nonexistent. The latter are distinguished externally from the females by small tufts of sensory hairs on their abdominal sternites. Forked fungus beetles use their horns in encounters between males—head-to-head pushing contests that last until one of the males retreats or is knocked off its fungal perch— and by single males to disrupt a copulating pair of fungus beetles. The

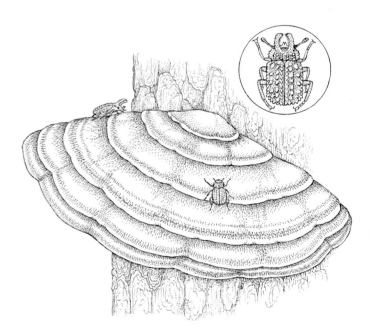

[ILLUSTRATION E]

The forked fungus beetle, *Bolitotherus cornutus*, is found in the forests of eastern North America. The forked horns of the males are used to compete for the favors of the female.

intruder may be successful in usurping the copulating male, but occasionally a third male will slip in and inseminate the female while the first two grapple with one another.

A series of experiments was designed to determine the importance of horn size in the forked fungus beetles. Males of equal body length and different horn size were chosen to participate in the study. Two males with different horn lengths were placed in a terrarium with a single female and a piece of fungus. They were observed for ten days. The males were then separated, and each was placed with a new female for another ten days, after which the males were reunited and placed with yet another female. Thus, each male fungus beetle was observed with a competitor present, alone, and again with his competitor. When the two beetles were confined together, the beetle with the larger horns successfully courted and copulated with the female 60 to 75 percent of the time. When the pair was separated, each male courted and successfully copulated with equal vigor and frequency, regardless of horn length. The fact that beetles with larger horns have greater access to females is consistent with field observations. Clearly, the major males can outcompete minor males, but do these results indicate that females prefer more well-endowed suitors?

In an experiment to test this idea, major and minor males were tethered in a terrarium so that they could move about but could not come in contact with one another. Females were courted equally by both males and mated with either one, demonstrating no preference for the size of the major male's horns. Although the female may evaluate several factors regarding her potential mate, horn size is apparently not one of them. Male aggression, not female choice, is what determines the greater success of the better-endowed fungus beetle.

What factors determine horn size in forked fungus beetles? If horn size is an inherited trait, then by virtue of the pressures of natural selection, one would expect to find a higher percentage of major males in the general population. Preliminary studies have shown that horn size is more likely to be determined by environmental rather than inherited factors. Females help determine the survivorship of the larvae and the size of the male horn when they select the site to lay their eggs. Female forked fungus beetles tend to be choosy in selecting egg-laying sites; thus the options within their immediate vicinity are limited. The abundance of minor males may reflect the relatively few high-quality egg-laying sites available to the females. In other words, even though a mother can influence the size of her son's horns and the vigor of her daughters, local environmental conditions limit her influence.

Horns are not always associated with sexual behavior. In a beetle closely related to the chrysomelid *Doryphora punctissima*, both sexes are armed with a spine located underneath the thorax. The males probably do not compete directly with one another through physical contests for females, but they do engage in pushing contests when placed together on small host plants. The attacking beetle attempts to dislodge its victim by working its horn under the edge of the opponent and prying it off the plant. Females also engage in this kind of scuffling to secure egg-laying sites on plants that are too small to accommodate more grazing activity than that of the offspring in a single brood.

Several genera of dung, earth-boring, and dynastine scarabs have a

bewildering array of horns on the head and/or back of the prothorax. Particularly impressive are members of the primarily Old World scarabaeine genera *Onthophagus* and *Proagoderus* and the neotropical athyreine genus *Athyreus*. The evolutionary process by which the tough chitinous exteriors of these beetles have extruded into these exquisitely developed armaments remains unknown. All of these species are thought to be adapted to very similar environments throughout their ranges.

One of the world's largest beetles is the elephant beetle, *Megasoma elephas*. [PLATE 29] The male bears a long, forward-projecting horn that has a smaller dorsal horn at its base, complemented by three smaller horns projecting from the front of the prothorax. Presumably male elephant beetles could use these horns in battle for the attention of females, but our observations with related horned scarabs, such as *Megasoma thersites*, *Megasoma cedrosa*, and *Dynastes granti*, suggest that the horns are used primarily to secure and defend as feeding sites the sap flows on tree limbs.

The disparate size between major and minor forms of the beetle *Podischnus agenor* suggests, at least in part, a genetic basis. Smaller males seem to compensate for their hereditary disadvantage by emerging from the egg before their larger, more aggressive brothers, and they disperse over much wider areas in search of females, thereby reducing the possibility of contact with major males. Having two game plans for bringing the sexes together helps ensure the survival of the overall population.

{THORACIC PARTS}

THE thorax is the powerhouse of the beetle body, enclosing an internal battery of muscles that drive the legs and wings. The three parts of the thorax—pro-, meso-, and metathorax—each bear a pair of legs, but only the meso- and metathorax may possess wings. The mesothorax usually bears a pair of thickened wings, the **elytra;** the metathorax supports a pair of sometimes intricately folded membranous wings. Each thoracic segment is then further subdivided into an upper sclerite, or **notum** (pro-, meso, or meta-, depending on the segment) and the ventral sternite.

One of the best-known modifications of the beetle thorax is that of the click beetles, or elaterids. [PLATE 30] Turned over on their backs, click beetles will right themselves by flipping up into the air with an audible click. The beetle accomplishes this feat by arching its back between the prothorax and the mesothorax, leaving only its prothorax and elytra in contact with the ground. By quickly contracting its ventral muscles, the stout prosternal process catches on the edge of a cavity in the mesosternum so that the muscle contraction is isometric. The muscle tension builds until suddenly the prosternal process snaps into the cavity, bringing the prothorax back in line with the mesothorax, jerking the beetle into the air. Although rather haphazard, this effort occasionally lands the click beetle on its feet, allowing it to scurry away. [ILLUSTRATION F]

GETTING A LEG UP
IN THE WORLD
EACH beetle leg has five distinct subdivisions: **coxa, trochanter, femur, tibia,** and **tarsus** in order from the body outward. In some beetles the tarsus is absent. The coxa connects the leg to the body in such a way that allows for horizontal, to-and-fro movement of the legs, enhancing the beetle's ability to move through or over its chosen medium—mud, sand,

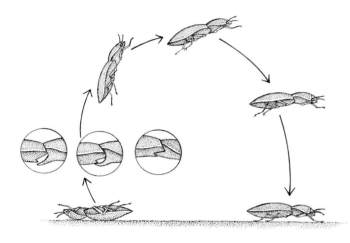

[ILLUSTRATION F]

To right themselves, click beetles of the family Elateridae will flip up into the air with an audible snap. Click beetles accomplish this feat by quickly contracting their ventral muscles, resulting in the stout prosternal process catching itself on the edge of a corresponding cavity on the mesosternum. Tension builds until the process snaps into the cavity, jerking the beetle into the air. Although rather haphazard, the beetle will eventually alight on its feet.

[PLATE 29]

The New World genus *Megasoma* contains some of the largest beetles in the world. One of these species, *Megasoma anubis* from southern Brazil, may be a pest of palm trees. The velvety appearance of this beetle is due to tightly packed hairlike setae. *Specimen courtesy of the Natural History Museum of Los Angeles County.*

[PLATE 30, *opposite*]

Click beetles, family Elateridae, are so named for their ability
to right themselves with an audible "click". Top:
Campsosternus leachi, Malaysia; bottom *Hemiops nigripes*, Java.
Specimens courtesy of Maxilla and Mandible.

[PLATE 31]

Male diving beetles, such as the North American *Cybister
fimbriolatus*, have their front tarsi modified to function as
suction cups which allow them firmly grasp the smooth and
slippery elytra of the female during copulation. *Specimen
courtesy of Maxilla and Mandible.*

decomposed organic matter, excrement, or water—by crawling, running, jumping, digging, or swimming. Simple, incised, or serrated claws located on the end of the tarsi enable beetles to perform amazing feats. Some chrysomelid beetles have oily pads on their tarsi that function as suction cups, making it nearly impossible for predators—and collectors—to extricate them from the surface of their host plants. The front tarsi of some male dytiscids are modified as suction cups, enabling these beetles to firmly grasp the slippery elytra of the female during copulation. [PLATE 31]

The front legs of the male harlequin beetle, *Acrocinus longimanus*, exhibit allometric development: The femur and tibia in larger beetles are extremely elongated. These elaborate outgrowths are presumed to be useful for moving about the branches that are the home of adult beetles. [PLATE 32] The long-armed chafer, *Euchirus*, and several large weevils also have greatly elongated front legs. [PLATES 33 AND 34] In the Kalahari Desert of Botswana, several genera of tenebrionid and anthicid beetles use their long spindly legs as stilts to put as much distance as possible between their bodies and the blistering hot sands. The legs of many beetles are armed with spines that lash out at their enemies. Others are flattened or fringed with setae, becoming oars that propel the beetle through water.

ON THE WING   INSECTS were the first animals to take to the air. They are the only truly winged animals, having evolved wings as distinct structures. This history is in direct contrast to that of pterosaurs, birds, and bats, all of which evolved wings from existing limbs. Although most insects have four functional flight wings, beetles have sacrificed the first pair, the elytra, to function primarily as protection for the delicate, membranous hind wings and soft abdomen underneath. [ILLUSTRATION G] The membranous hind wings are supported by a framework of veins that may double as hinges along which the wings can fold. Females of the Mediterranean scarab *Pachypus* lack both membranous hind wings and elytra, having instead strongly thickened abdominal segments that perform the protective function of the elytra.

Many wingless beetles, especially those that are highly mobile and adapted to living in hot deserts, have a dead air space between their elytra and abdomen to insulate the body from sudden changes in temperature. A classic example of a beetle with this subelytral air space is the North American meloid *Cysteodemus armatus*, which can be seen running quickly about the creosote flats of the Southwest on warm spring days. Studies of beetles such as the armored darkling beetle, *Eleodes armata*, have shown that these subelytral cavities may be warmer than the ambient temperature, indicating that they may function as both convective cooling and heat buffering systems.

{THE ABDOMEN}

THE beetle abdomen houses the bulk of the circulatory, respiratory, digestive, excretory, and reproductive systems. Generally, the dorsal portion of the abdominal segments, the tergites, covered by the elytra is only lightly sclerotized and very flexible. In the staphylinids and many nitidulids, however, which have short elytra, the tergites are more heavily sclerotized.

The terminal segments of the abdomen have been variously modified

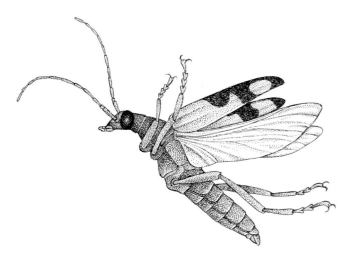

[ILLUSTRATION G]

Insects were the first animals to take flight and are the only truly winged animals, having evolved wings as distinct structures. While most flying insects possess four functional wings, beetles, such as this long-horned wood boring beetle, *Strangalia soror,* have sacrificed the first pair to function primarily as protection for the delicate hind wings and abdomen. The clawed legs are extended like grappling hooks in anticipation of landing.

[PLATE 32]

The front legs of the male *Acrocinus longimanus*, South America, and the hind legs of *Sagra* sp., Malaysia, are greatly exaggerated in their development. *Specimens courtesy of Maxilla and Mandible.*

[PLATE 33, *opposite*]

The male euchirine scarab, *Euchirus longimanus*, with its
elongate front legs armed with spines, is easily distinguished
from the female, which possesses normally developed legs.
Specimens courtesy of Paul McGray.

[PLATE 34]

The female *Cheirotonus parryi*, a large euchirine scarab unique
to Malaysia. *Photograph by Charles Bellamy.*

to facilitate egg laying by the female and insemination by the male. In females, an internal elongate **ovipositor**, the specialized organ for egg laying, is characteristic of many beetle families that complete their life cycles in wood. Species adapted to laying their eggs directly on the surface of soil have short, stout ovipositors. Additional modifications reflect modes of laying eggs within plant tissues or on other exposed surfaces. The terminal segments of the gyrinid abdomen, which are laterally compressed and highly mobile, serve as a rudder in their travels across ponds and streams.

One of the most bizarre modifications of the beetle abdomen is known as physogastry. **Physogastry** is a condition in which the abdomen is enlarged, usually through increased growth of the connective membranes, expansion of the internal fat bodies, and subsequent secondary sclerotization of the abdominal sclerites. Physogastry is usually accompanied by the presence of glands and pores that exude substances that seem to enhance positive interactions between termitophilous (termite-loving) beetles and their termite hosts. Completely adapted to the environmental conditions of their hosts, physogastric beetles often die shortly after being removed from the confines of their nest.

Physogastry has evolved independently several times in termitophilous staphylinids and corythoderine scarabs. In staphylinids physogastry may be induced by the presence of some unknown substance in the termite nest. During growth of the membranes between segments the ventral membranes expand more quickly than the dorsal segments, resulting in the abdomen becoming recurved over the back of the beetle. As the abdomen increases in size, the morphological anterior (functional posterior) of the abdomen becomes secondarily sclerotized. The legs also become thicker and more sclerotized, possibly to support the increased weight of the abdomen. The resultant beetle bears an uncanny resemblance to a termite. Even though the visual resemblance is lost on the blind termites, the physogastric structure of the abdomen provides sufficient tactile cues for the termite to accept these termitophilous beetles into the colony as one of their own.

{INTERNAL SUPPORT}

THE exoskeleton is characterized internally by a series of infoldings, ingrowths known as **apophyses,** struts, and ridges that form the internal framework from which muscles exert leverage on other sternites and appendages. The degree of musculature directly reflects the development and extent of use of structures associated with the head, thorax, and abdomen. Adult and larval mandibles derive their leverage through the interaction of sturdy internal apophyses and powerful muscles. The musculature of the thorax drives the organs of locomotion, the wings and legs. The amount of supporting musculature directly reflects the use or loss of wings. The abdominal musculature supports the various organ systems, and its diversity reflects the adaptations of the adults and larvae for ventilating movements, egg laying, extending of the genitals, folding and unfolding of the wings, and movements designed to right the body. The more fused the sternites of the abdomen are, the less mobile is the adult beetle. Such reduced mobility is accompanied by a reduction or loss of ventral longitudinal muscles.

This internal framework of supporting structures and muscles supports

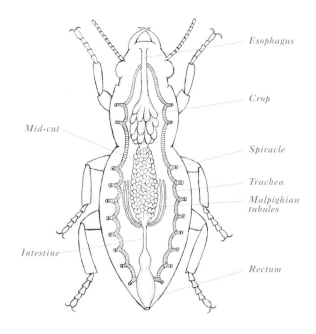

Esophagus

Crop

Mid-cut

Spiracle

Trachea

Malpighian
tubules

Intestine

Rectum

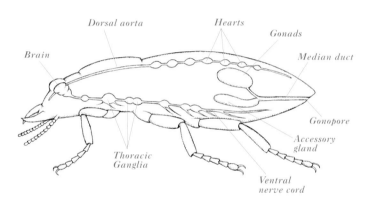

Dorsal aorta   Hearts

Gonads

Brain

Median duct

Gonopore

Accessory
gland

Thoracic
Ganglia

Ventral
nerve cord

[ILLUSTRATIONS H AND I]

Beetles are filled with complex organ systems which faciliate
their ability to breathe, metabolize food, circulate blood,
reproduce and sense their environment. Top: respiratory and
digestive systems; bottom: circulatory, reproductive,
nervous systems.

a series of complex organ systems, providing beetles with the capacity to do what vertebrates do: sense their environment, digest their food, excrete waste, circulate their blood, breathe, and reproduce. [ILLUSTRATIONS H AND I] But this is where the similarity ends. With their dorsal blood vessel (see "Circulation") and paired ventral nerve cord (see "Sensation"), beetles are vastly different from vertebrates. Compared in this way to vertebrates, beetles can be said to be built inside out and upside down.

{ALL SYSTEMS GO!}

DIGESTION THE digestive system is one of the most conspicuous internal systems of beetles. Once the mandibles and maxillae, the external organs of ingestion, have lacerated, shredded, pulverized, or imbibed the seemingly limitless kinds of solid and liquid foods consumed by beetles, it is the task of the digestive system to process and metabolize the food. The digestive system is not unlike a conveyor belt, moving materials internally from the mouth to the anus. The beginning of the digestive tract is connected to the head by musculature, and the remainder is supported by, if anything, tracheal branches (breathing tubes).

All three regions of the digestive tract (**foregut**, **midgut**, and **hindgut**) exhibit some degree of **peristalsis**, the rhythmic muscle contractions that push materials along the tract. Along the way water may be extracted or conserved and nutrients absorbed by the body. Solid materials may be further reduced in size by a battery of grinding teeth located in a region at the rear of the foregut known as the **proventriculus.** Digestive enzymes secreted by the salivary glands and from the midgut further reduce ingested foods into their smallest components in preparation for being absorbed. The hindgut is the primary source of absorption of water and dissolved nutrients. Materials moving through this region become progressively drier until they reach the **rectum,** where they are treated and converted into a fecal pellet (the polite term for solid beetle waste is **frass**) before it is discharged through the **anus.** But "waste not, want not" appears to be the motto of the larvae of many cetoniine scarabs, which construct a protective pupal chamber from their fecal pellets.

EXCRETION IN addition to the waste that is generated and eliminated by the digestive tract, waste products must be removed from the blood. Beetles remove blood wastes via **Malpighian tubules,** named by the German anatomist Johann Friedrich Meckel in 1829 after Marcello Malpighi, the Italian anatomist who first described these organs in silkworms, calling them *vasa varicosa*, in 1669. The Malpighian tubules, which are attached to the hindgut, extract nitrogenous wastes (that is, wastes that contain nitrogen), such as uric acid and salts, from the blood and dumps them into the hindgut. Much of the remaining water and some of the salts are then reabsorbed by the body, leaving uric acid crystals to be discharged with the solid waste.

REPRODUCTION THE tremendous reproductive capacity of beetles is at the very core of their success and has permitted them to radiate out into nearly all elements of the environment. Almost all beetles are either male or female. Only rare exceptions are both sexes simultaneously; these rare forms are called **gynandromorphs**. In the most conspicuous cases of gynandromorphy, one side of the beetle is structurally male, the other side female. Lucanids, dynastine scarabs, and cerambycids provide some of the most dramatic examples of this abnormality.

In some beetles, most notably some weevils, males are rare or completely lacking; the females of these species are referred to as **parthenogenetic**. Parthenogenetic females reproduce without the help of males. The advantage of single-sex reproduction is that time is not lost in seeking a mate and on the ensuing courtship. However, although sex may seem like a waste of time, it does provide a useful mix of necessary ingredients to ensure variability among offspring, which improves the chances of survival for most species.

Typically, each sex is equipped with its own set of reproductive organs, consisting of paired gonads (sexual organs) connected to a median duct that leads to an external opening called the **gonopore**. Accessory glands secrete materials that in the male are used in the production of the **spermatophore**, a capsule that protects the sperm, and in the female contribute to adhesive or protective coatings for the eggs.

The female reproductive organs consist of a pair of **ovaries** (or, in the case of the dung scarabs, a single ovary on the left side), each with one or more subcompartments called **ovarioles**. Short-lived beetles generally tend to have more ovarioles, thus increasing the number of eggs available at any given time and thus maximizing their reproductive potential. Beetles that live longer are not under as much pressure to produce all their eggs at once and can afford to have fewer ovarioles. Each ovary bears an **oviduct**, a tube that leads into a common median oviduct. Attached to the median oviduct is the **spermatheca**, a sac that stores live sperm from the male, in some species for many months. The eggs are generally fertilized while being laid, as they pass through the median oviduct. The median oviduct or other glands may secrete sticky substances, causing the eggs to adhere to one another and providing a protective coating. In cassidine chrysomelids and most hydrophiloids, these glands produce a threadlike material that is secreted in a manner to resemble a cocoon. The female of the buprestid *Steraspis speciosa* deposits her eggs wrapped in a protective case, or **ootheca**. The source of this coating, however, has not been linked to specialized secretory glands.

The male reproductive system roughly mirrors that of the female, consisting of a pair of **testes** whose ducts, the **vasa deferentia** (singular *vas deferens*) join in the **ejaculatory duct**. The testes of all Adephaga, as well as some members of the Archostemata and Myxophaga, are unusual in that they consist of a single long tube tightly coiled and wrapped in a membrane. Accessory glands attached to the upper end of the ejaculatory duct provide materials for the production of spermatophores in some beetles.

The genitalia of male beetles help identify different species. The diversity and complexity of the genitalia suggests that they have been selected for reasons other than just the transfer of reproductive materials. These reproductive organs are often characterized by variously placed hooks and spines, which may assist the male in his efforts to grasp and remain coupled with the female. It has been suggested that the unique development of male genitalia functions as one of many isolating mechanisms, a type of "lock-and-key" system that allows coupling only with the corresponding female apparatus. This and other exclusionary systems prevent beetles from wasting precious time and energy trying to mate with the wrong species. More likely, however, the interlocking characteristics of the male and female genitalia serve to reduce competition between males of the same species.

The primary function of the reproductive organs is to ensure not the survival of the individual, but rather the success of future generations of beetles. Unlike vertebrate systems, insect reproductive systems show no evidence of having a regulatory effect on the activities of other systems or on the development of secondary sexual characters, such as horns.

CIRCULATION   DESPITE the myriad of beetle habits, the circulatory system is remarkably uniform throughout the Coleoptera. Unlike the closed system of vertebrates, in which the blood remains inside arteries and veins, beetles possess an open circulatory system that consists primarily of a dorsal aorta with a series of pumps, or hearts, located posteriorly. From five to eight openings called **ostia** allow blood to flow into the heart. Paired **alary muscles**, which drive the pumping action of the hearts, insert on either side of the heart chambers, originating on the sides of the abdominal tergites. The **pericardial membrane** surrounds the heart, separating it from the other internal organs. Extending from the heart to the head, the **aorta**, the vessel that carries blood through the body, opens just behind the brain. Large beetles, such as *Dynastes*, are reported to have branches arising from the aorta into the thorax.

The blood of the open circulatory system does not carry oxygen, but it does transport nutrients and hormones. The pumping action of the multiple hearts forces nutrient-rich blood into the head region, where it spills out into the various body cavities, bathing the tissues. Eventually the blood travels back into the abdomen, where it enters the system through the ostia to be pumped forward again by the hearts.

SENSATION   THE nervous system consists of a ventral **nerve cord**, with paired **ganglia** (bundles of nerve cells), that conducts the nerve impulses required to innervate (stimulate to activity) all parts of the beetle body. The nervous system of the adult beetle is more specialized than that of the larva, which retains a primitive system consisting of three thoracic and eight abdominal ganglia. The brain (also referred to as the dorsal ganglion), which is encapsulated by the head, is divided into three sections. The anterior **protocerebrum** processes images coming in through the optic nerve (the nerve from the eyes). The **deuterocerebrum** innervates the antennae. The paired **tritocerebral lobes** are connected beneath the esophagus and innervate the labrum and the foregut. Following the tritocerebral lobes are the subesophageal and thoracic ganglia. The subesophageal ganglion innervates the remaining mouthparts, while the thoracic ganglia process the nerve impulses required to drive the legs and wings. The distinct, mobile nature of the pronotum in most beetles is mirrored internally by a large, well-separated prothoracic ganglion. The remaining two thoracic ganglia may be more or less fused in some beetles, but if not, they are clearly joined by connective tissue. The first abdominal ganglion is usually fused to the metathoracic ganglion, and the posteriormost two abdominal ganglia are often fused.

BEETLES have a very effective respiratory system, which enables them to survive in a variety of terrestrial and aquatic habitats. The movement of oxygen and carbon dioxide depends, for the most part, on passive transport, or simple diffusion. Dependence on such a system has imposed a severe constraint on beetles—the limitation of size—because diffusion is effective only over relatively short distances. However, the size constraint has not been an impediment to the overall success of beetles, for their relatively diminutive size has allowed them to exploit a multitude of niches not available to or only poorly utilized by larger vertebrates.

The respiratory system consists of a series of tubes, called **tracheae**, and air sacs. At the end of each trachea is a **spiracle**, a valve on the surface of the exoskeleton through which the trachea draws air. The known total volume of the tracheal system in beetles ranges from 20 to 40 percent of the body volume. The air sacs form at the blind ends of tracheae or by dilations along the main tracheal trunks and help keep the body light and dissipate heat. These sacs also improve tracheal ventilation by supplementing the opening and closing of the spiracles, and they help regulate blood pressure with the aid of abdominal movement.

The tracheae are infoldings of cuticle that are shed with the rest of the exoskeleton when the beetle larva molts. The tracheae are supported externally by a series of coils, called **taenidia** (which the air sacs lack), much like the plastic air tubing used with old portable hair dryers. The tracheae branch into progressively smaller tubes until they come into direct contact with the cells of the muscles and internal organs. Oxygen, carbon dioxide, and other gases are exchanged between these cells and the external atmosphere following the general rule of diffusion: Gaseous molecules travel from areas of high concentration to areas of low concentration.

Some the most intriguing examples of beetle respiration are observed in aquatic beetles, such as the dytiscids, hydrophilids, dryopids, and elmids. Hydrophilids draw a layer of air over the ventral surface of their abdomen. Dytiscids trap air beneath their elytra. Dryopids and elmids use a **plastron**, a film of air that envelops their entire body and is held in place by a thick layer of water-repellent setae. These trapped bubbles of air function as a physical gill. Dissolved oxygen in the water is at a higher concentration than that contained within the bubble maintained by the beetle. The partial pressure of oxygen within the bubble falls below that of the water as the oxygen is used by the beetle, causing more dissolved oxygen in the surrounding water to diffuse into the bubble. Aquatic beetles obtain more oxygen through this temporary structure than was available in the original volume of air that was trapped. The presence of nitrogen, a nonrespiratory gas, is essential for the bubble to act as a physical gill. Without it, the partial pressure of oxygen does not change as it is used. The need to replenish its supply of nitrogen is what forces the beetle to return to the surface and trap another air bubble. This self-contained underwater breathing apparatus may also function in a manner similar to that of the swim bladder of fish, controlling the beetle's buoyancy in water.

{PARTS IS PARTS—OR ARE THEY?}

EACH beetle resembles a unique machine, its many parts constructed according to specifications determined by dynamic environmental

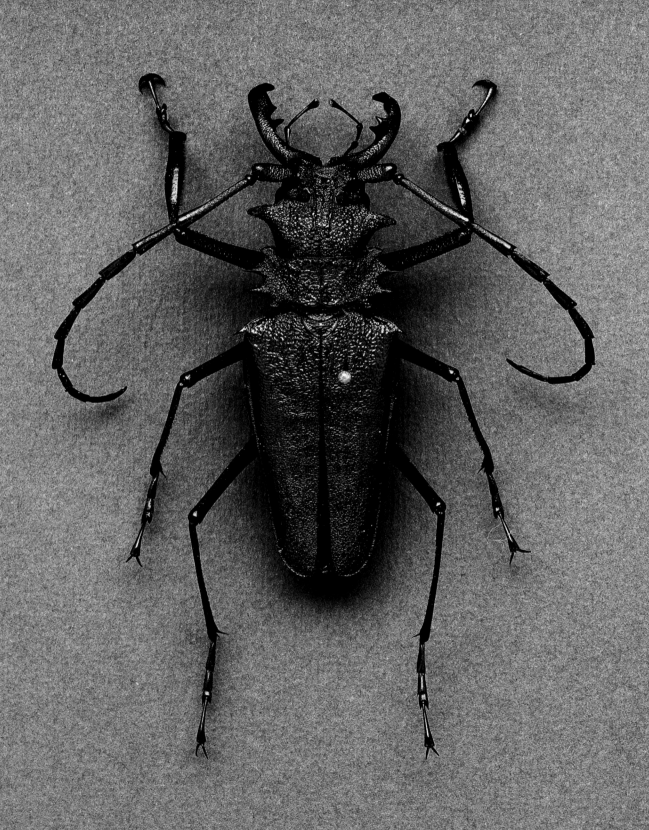

parameters and thoroughly tested during millions of years of fits and starts. [PLATE 35] As beetles evolve, their structures become modified—so much so that their origins may become obscured. However, because the same features do not generally change at the same rate, if at all, in all populations, today we have the opportunity to observe, by carefully analyzing beetle morphology, a mosaic that reveals a discernible pattern of structural modifications that reflect their evolutionary legacy.

Despite numerous examples to the contrary, the basic blueprint for what we call a beetle demonstrates an overall economy of form and function. Yet the morphological data reflects only part of the beetle's machinery. Coleopterists must remember that beetle morphology is merely the outward manifestation of genetics, physiology, and behavior, formed within a tapestry of complex interactions with other organisms, atmosphere, oceans, and soil.

The Cartesian approach to science, developed by the French mathematician and philosopher René Descartes, is that if we view the world as a machine, we can grasp the underlying principles by which it is governed by disassembling the machine and studying its parts. This method of study has served us well in coleopterology, at least from the viewpoint of advancing our knowledge of beetle phylogenetics and in our efforts to define species for cataloguing the fauna of different ecosystems. But viewing beetles simply as machines, without understanding their role in the ecosystem, is a narrow perspective that reflects intellectual, spatial, and temporal limitations. As the world's ecosystems continue to shrink in the wake of human exploitation—a direct result of our ever-burgeoning population—our approach to all the sciences must continue to evolve from an analysis of parts to a necessarily more holistic approach. We must learn to view beetles not as machines, but as conduits of energy flowing through the entire biosphere.

three

THE BEETLES—YESTERDAY AND TODAY

*Judging from their structure, habits, and economy, there are
reasons why beetles ought to excel every other class of organized
beings as exponents of the past geography of the globe.*

•

Andrew Murray (19th-century British coleopterist)
The Journal of the Linnean Society, 1870

BY virtue of their marvelously adapted morphology, beetles have clearly demonstrated a propensity for colonizing new habitats and exploiting underutilized niches. Generally speaking, **niches** are modes of food acquisition or preferred food sources. **Habitats** are sites where these modes of food acquisition take place. The broader the niche and habitat, the greater ability a beetle species has to adapt to change and the greater chance for continued matings to enhance genetic flow. Employing a staggering array of behaviors, some of which we present in this chapter, beetles are ideally suited to meet the rigors of their demanding and unforgiving environments.

{HOME IS WHERE THE BEETLE IS}

BEETLES have been so successful in large part because of their small size, which allows them to occupy, exploit, and diversify within countless niches in all kinds of habitats. Beetles inhabit all possible environments—from beaches to rocky fell-fields on mountain peaks thousands of meters above sea level to arid deserts to tropical rain forests.

One general rule about the diversity and distribution of organisms is that species diversity decreases toward the poles of Earth. In temperate regions, what beetles lack in species diversity they make up for in sheer numbers; populations are often large and wide-ranging. In equatorial regions, the converse is true: The number of species is large and diverse, but each is represented by a smaller, more localized population. [PLATE 36] Seasonality is a primary factor that influences the diversity and population size of beetles throughout the world. In temperate and polar regions, seasonal fluctuations of climate are predictable. Periods of warmth alternate with periods of cold, precipitation with drought, and long days with short

[PLATE 36]

Goliathine scarab beetles of the subfamily Cetoniinae, which are tropical in their distribution, are often brilliantly colored. The males may be distinguished by sometimes elaborate armatures arising from the head. Top to bottom, left to right: *Dicranocephalus wallichi*, India; *Eudicella gralli*, Central Africa; *Fornasinius russus*, Central Africa; *Rhamphorrina splendens*, Zimbabwe; *Fornasinius russus*, Central Africa; *Mecynorrhina torquata*, Central Africa; *Dicronorhina derbyana*, South Africa. *Specimens courtesy by Maxilla and Mandible.*

days. In tropical regions, these climatic fluctuations are less seasonal, and at the equator seasonal variations are barely perceptible. The standing body temperature of beetles depends directly on the temperature of the surrounding environment and directly limits the occurrence of most species to within very specific parameters. Abiotic factors, particularly the availability of water, also limit beetle distribution. Many other biotic and abiotic factors combine their influences, resulting in one of the most distinctive features of beetles, their natural distribution.

Some species of beetles are restricted to the tropics; others are confined to temperate forests, deserts, or aquatic environments. Still other species, although once broadly distributed, have become taxonomic or ecological relicts, restricted to mountaintops and other disjunct habitats, exhibiting distributions not unlike species of beetles that inhabit islands. Beetles that feed on plants or parasitize other organisms are necessarily limited by the distribution of their hosts, but rarely does the distribution of the beetles coincide fully with that of the host. Another critical factor that limits beetle distributions is the presence or absence of organisms that make hosts or prey of the beetles themselves. When beetles are moved from the areas of their natural distribution to other habitats, the lack of checks and balances in the new environment, which in their home habitat would have been provided by natural predators, often allows their populations to grow unchecked. Thus, they may become pests.

To study beetle populations, we collate data from specimens in various collections into detailed distribution maps. Knowing the distributions of beetles, along with those of other animals and plants, has provided the foundation for proposed schemes by which we can divide the planet into zones characterized by particular flora and fauna. These zones are referred to as **biogeographic regions**. In 1870 Andrew Murray proposed that beetles could be divided into two groups according to distribution: those found predominantly in temperate regions and those found in warmer tropical climates. Alfred Russel Wallace, working with a variety of organisms, later proposed the scheme of biogeographic regions used today.

To date, discussions of worldwide beetle distributions are usually restricted to families or lower levels of taxa. Most papers that deal with larger taxa focus on a particular biogeographic region. Worldwide reviews, on the other hand, tend to cover taxa with a relatively small number of species. Although most of these distributional studies are primarily descriptive, some works attempt to analyze both present and past distributions on the basis of available fossil evidence to propose centers of origin for taxa and methods of dispersal based on distribution. These studies are even more compelling when compared to the distributions of other organisms, to past and present geography, and to phylogenetic history.

INTO THE PAST   TO set the stage for a discussion of where and how modern beetles go about their lives, we will briefly look at their evolutionary history, as interpreted through the examination of fossil evidence. Beetle fossils are abundant and their persistence is due to one of the most conspicuous features they exhibit during life, their tough exoskeletons. The sclerites of the head, thorax, and elytra of ancient beetles frequently are preserved.

The Coleoptera probably arose in the Permian period (230 million

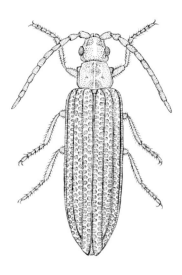

years before present) from ancestors similar in appearance to the modern insect order of Megaloptera, which includes the alderflies, dobsonflies, and fish flies, whose larvae are aquatic. Remains of fossilized "protocoleopterans" from the Lower Permian resemble present-day beetles of the family Cupedidae, whose elytra have distinct ribbing and sculpturing. [ILLUSTRATION J] The somewhat flattened form of these and other species of fossil Protocoleoptera suggests that they occupied spaces under the loose bark of trees, and thus that this habit was ancestral in the order. Protocoleopterans differ from modern beetles in that their elytra are less regularly sculptured, are not closely fitted to the abdomen, and extend beyond the abdomen. Some protocoleopteran fossil specimens bear antennae that have thirteen segments, whereas the basic antennal complement in modern beetles is eleven segments.

During the Mesozoic era, protocoleopterans all but disappeared, replaced by the precursors to modern beetles. By the middle of the Triassic period, all four of today's suborders of beetles (Archostemata, Adephaga, Myxophaga, and Polyphaga) are known to have existed. Before the end of the Jurassic period the lineages of all of the modern superfamilies of Coleoptera (see Appendix 1) were established. During the Jurassic, deposits of resin secreted by conifers began to appear, suggesting that these ancient trees were attempting to protect themselves from the attack of wood-boring insects similar to the modern bark beetles of the family Curculionidae. The Cretaceous period provided beetles with an evolutionary fast track resulting from the sudden appearance of flowering plants, the angiosperms. During this period we encounter the first beetles trapped in **amber** (fossilized resin). [PLATES 37–44]

Amber has long fascinated humans. The Romans called it *succinum*, meaning literally "sap stone"; to the Greeks it was *elektron* (the root of the English word *electricity*) because static electricity could be created by rubbing amber. The word *amber* is derived from the Arabic *anbar*, which refers to both fossilized resin and whales. Although found throughout the world, amber is best known from deposits in the Baltic region, the Dominican Republic, Mexico, China, Burma, Sicily, Canada, and Siberia. In the United States, amber deposits are found in Alaska and New Jersey.

Beetles preserved in amber deposits can be reliably identified. At least sixty families of beetles are represented in amber-preserved forms, and most of these species can be attributed to existing tribes and genera. A unique feature of amber is that it preserves organisms in habitats that are generally poorly represented in the fossil record, such as those thought to have existed in ancient rain forests. By comparing fossil species with their modern relatives, biogeographers are able to make inferences concerning the ancient habitat of the entombed beetles. Further, beetles in amber have provided insights into speciation (the development of new species) and extinction rates with data revealing the longevity of taxa. Fossil beetles also reveal the age of their symbiotic relationships with other organisms, particularly other animals.

The high degree of tissue preservation in beetles preserved by amber has attracted the attention of physiologists and molecular biologists. Muscle fibers and cell structure are preserved in amazing detail, but it remains doubtful that complete sequences of DNA will be able to be recovered from cells preserved in amber. Nevertheless, even small portions of this pre-

[PLATES 37, 38, 39, 40, 41, 42, 43, AND 44]

At least 60 families of beetles have been preserved in amber, or fossilized resin. The following examples are all from the Dominican Republic: Ciidae; larval Brachyspectridae; Rhipiphoridae; Cerambycidae; paussine carabid; Curculionidae; Lampyridae. *Photographs by George Poinar.*

served DNA may yield fascinating results when compared to the DNA of closely related existing species, providing firsthand information about the extent to which organisms have evolved in a given period of time. Assessing the similarities and differences of the DNA can help us determine the line of evolutionary descent of beetles.

Most of the fossil beetles from the Quaternary period (1.6 million years before present), particularly those whose remains were deposited in the last 500,000 years, are identical to our modern fauna. These chitinous bits of our recent past—for the tissues of the exoskeleton have not been replaced by mineral deposits—are surrounded by permanently frozen debris, pressed into water-laid sediment, sheltered among prehistoric mammalian dung heaps, or steeped in asphalt.

The famous La Brea Tar Pits of southern California, located only minutes from downtown Los Angeles, and another site to the north in nearby McKittrick, represent two of the richest deposits of Late Pleistocene mammal and bird fossils in the world. These deposits are also incredibly rich in insect fossils that are encased and preserved in sticky asphalt. The asphalt was forced through fissures in the Earth's surface by methane gas, where it formed shallow sticky pools in low-lying areas. These pools were often covered by water or plant debris, which led to the entrapment of thousands of animals. Carrion and dung beetles, among others attracted to the bodies of dead or dying animals, were also trapped and preserved in the sticky stuff. Early paleoentomologists, scientists who study fossil insects, believed that all fossil species represent extinct species or subspecies of existing forms. In fact, however, most fossils represent modern species.

Two species of presumably extinct scarabs were described from the La Brea deposit, *Onthophagus everestae* and *Copris pristinus*, with their nearest existing relatives residing in Texas and Mexico, respectively. The disappearance of mammoths, sloths, camels, and other large herbivores no doubt reduced the availability of dung, ultimately leading to the extinction of these and other beetles who, in one way or another, depended on these mammals for their living.

We base our interpretation of beetle fossils on inferences taken directly from our knowledge of the distribution and habits of modern fauna. We assume, particularly in recent fossil assemblages, that ancient beetles had environmental requirements similar or identical to those of their modern counterparts. With this knowledge scientists can reconstruct paleoclimates from assemblages of beetle fossils with some degree of confidence. However, since rapid changes in past climates could have thrown the beetle fauna into a state of flux, resulting in unique, short-lived mixtures of species for which there is no comparison among modern beetle assemblages, scientists must be cautious with their analyses. Clearly, our understanding of modern beetle distributions strongly influences our perceptions of beetles past and present.

HOTHOUSE DENIZENS PACKED with a vast spectrum of life spanning all the kingdoms of life, from slime molds and fungi to towering trees and the entire range of animals, rain forests are home to equally diverse sets of beetles, from species that dwell in soil to those that spend their entire life among the leaves and flowers of the canopy, hundreds of meters above the forest floor. Recently our colleague

Geoff Williams published a book about the "hidden" rain forests in Australia, the culmination of studies to determine how rich the diversity of beetles can be in even a small rain forest. Working with a patchwork of relict forests in the Manning River area of New South Wales, Williams found seventy-one families of beetles represented by more than 700 species!

The climatic conditions of rain forests are nearly uniform; temperature and humidity both remain consistently high. The climatic constancy of tropical rain forests allows the size of beetles, among other features, to increase well beyond that of their counterparts in temperate regions. With moisture and heat loss much less of a problem, all structures and dimensions can swell and elongate—and they do. The largest beetles in the world live in tropical regions: the African Goliath beetles, *Goliathus;* [PLATE 45] the elephant beetles, *Megasoma*, of tropical Central and South America; the atlas beetles, *Chalcosoma*, of Southeast Asia; and the giant cerambycid *Titanus giganteus*, of northern South America.

RESOURCE ECONOMISTS

AT first glance, the blistering deserts of the world seem to be the antithesis of rain forests. Receiving little rain and exposed to sometimes wildly oscillating fluctuations in temperature, deserts are home to a rich and dominant beetle fauna whose basic physical features are remarkably similar throughout the world. The parallel evolution reflected in this similarity is a result of the fact that beetles living in the Sahara and Kalahari deserts of Africa, the Gobi Desert of China, the Atacama of South America, the Mojave and Sonoran deserts of North America, or the great central deserts of Australia have all adapted both behaviorally and morphologically to cope with the lack of water in these similarly harsh environments.

Beetles have developed many adaptations for conserving water in the inhospitable, arid desert environment: (1) Thickened exoskeletons with fused sclerites and waxy coatings helping the beetle to keep water in. (2) The beetle can voluntarily close its spiracles to reduce the amount of moisture lost through respiration. (3) Increased insulation afforded by a coating of hairlike setae, called **pubescence**, helps the beetle retain water. (4) Powdery secretion of a waxy coating called pulverulence helps prevent water loss. [PLATE 46] In the Algodones sand hills of southeastern California, the buprestid *Hippomelas dianae*, a greenish bronze jewel beetle, is dressed with longitudinal bands of whitish or yellowish pulverulence on its upper surface. This waxy covering not only assists the beetles in its efforts to conserve water; it also helps camouflage the insect from sharp-sighted predators.

Most desert-dwelling species of the family Tenebrionidae are wonderfully adapted to desert life. [PLATE 47] These wingless and usually black species escape extreme temperatures by remaining buried in the sand during the heat of the day, where even a few inches below the surface the moisture significantly lowers the temperature. [ILLUSTRATION K] As the sun begins to descend to the horizon, the beetles emerge, often in large numbers, to forage and seek mates. Many of these tenebrionids, such as *Cardiosis moufleti*, superficially resemble the streamlined aquatic diving beetles of the family Dytiscidae, with their bodies compressed top to bottom, allowing them to dive quickly into the sand in search of cooler surroundings.

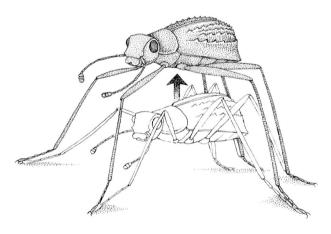

[ILLUSTRATION K]

The black and wingless desert dwelling tenebrionids of southern Africa have evolved several strategies for dealing with extreme temperatures. The Namibian *Onymachris* raises itself on stilt-like legs to distance itself from the blistering sands.

[PLATE 45]

The African Goliath beetles were named after the giant vanquished by David in the Old Testament. Adult beetles may be found feeding on sap exuding from the wounds of certain trees. Top: *Goliathus goliatus*, equatorial Africa; top: *Goliathus meleagris*, Congo; bottom: *Goliathus goliatus*, equatorial Africa. *Specimens courtesy by Maxilla and Mandible.*

[PLATE 46, *above*]

The powdery waxy coating, or pulverulence, secreted by the exoskeleton of some beetles, helps to prevent water loss. The small jewel beetle, *Lepismadora algodones*, endemic to the Algodones Sand Hills of southern California, exhibits this water saving feature. *Photograph by Charles Bellamy.*

[PLATE 47]

The thickened exoskeletons of many tenebs, such as *Akis reflexa* from Spain, acts as a deep sea diving suit in reverse, helping them to retain moisture in environments where water is scarce. *Specimen courtesy of Maxilla and Mandible.*

BY looking at the distribution of island life we can make many predictions about the biology of more broadly distributed terrestrial organisms, especially in terms of colonization, succession, and extinction. Islands of continental origin lie just off the coasts of continents, to which they were once connected. The beetle fauna of continental islands tends to mirror that of the continental fauna and has relatively few **endemics**, species whose distributions are restricted to the island. Oceanic islands, volcanic in origin, have never had a continental connection and tend to harbor more endemic species of beetles than islands of continental origin have. The oceanic island fauna is said to be disharmonic—that is, many of the taxa expected to exist there do not, primarily because they have simply failed to reach these remote pieces of terra firma.

Beetles may colonize both continental and oceanic islands by active means of transportation, such as flying. Passive transport, such as rafting on logs, may assist the movement of many wood-boring species and is primarily responsible for the dispersal of these beetles throughout the islands of the Caribbean. Larger species, such as the jewel beetle *Merimna atrata* or nocturnal species attracted to the lights of seafaring vessels, may stow away aboard ships and be carried considerable distances outside of their natural distribution. Smaller beetle species may be carried vast distances along wind currents generated by fierce storms and become established in new areas, assuming that they are fortunate enough to alight upon a suitable landfall rather than slipping into the depths of the open ocean.

In addition to true islands, some isolated terrestrial areas, such as the forest canopy or mountaintops, are, in effect, islands. These areas are frequently occupied by endemic beetles. For example, the broad expanse of the Sonoran and Chihuahuan deserts, which stretch across the southwestern United States and northern Mexico, are punctuated by a series of mountain ranges, each resembling an island of emerald green surrounded by a sea of desert. These mountain retreats are havens for beetles of the genus *Plusiotis*, whose species are not well adapted to the harsh environments of the lower elevations and would soon perish if they were forced out of their preferred forests of oak, juniper, pine, and fir. In the Santa Rita Mountains southeast of Tucson, Arizona, three species of this beautiful scarab genus are known to occur. *Plusiotis gloriosa* is found on juniper and is collected in the lower elevations of the mountains. *Plusiotis beyeri* prefers instead oak trees at slightly higher elevations. [PLATE 48] A third species, *Plusiotis lecontei*, feeds on pine and is distributed at even higher elevations. Despite this apparent ecological separation, by carefully selecting a habitat with a mixture of all the host tree species, collectors can obtain specimens of all three species at the same locale.

Isolated mountain peaks may also be considered "island" habitats. Beetles who reside at high altitudes, above the timber line, must deal with climatic and/or physiological factors that are not imposed on species that inhabit lower elevations. Some such factors are increased ultraviolet radiation, less oxygen, fewer available niches, and longer periods of cold. These harsh conditions require adaptations that tend to select against diversity.

Predatory carabids have been recorded at elevations of up to 5000 meters in the Himalayas. One species has been found hibernating under 10 meters of snow! Certain species of dytiscid water beetles, whose larvae and

[PLATE 48]

In the Santa Rita Mountains of southeastern Arizona the ruteline scarab *Plusiotis beyeri* rests on its host, the oak.
Photograph by Charles Bellamy.

adults are aquatic, do not exist below certain elevations and are found only in mountain lakes and ponds created from melting snow. The adults hibernate under stones, often under snow cover, near the edge of a pond or lake, to avoid being frozen in water. Carabids, rove beetles, and weevils that live at high elevations are often adapted to the exposed windy conditions of their montane habitats by the reduction or loss of their wings. These species frequently live near the edge of melting snow, in organic debris or under moss and rocks, foraging for food on the snow surface only during the warmest parts of the day.

TROGLODYTES — LIKE true islands and mountaintops that harbor endemic populations of beetles, caves are isolated environments. These subterranean habitats are characterized by relatively constant temperatures comparable to the average temperatures on the surface, and often with continually wet surfaces. The atmosphere is humid, the level of carbon dioxide is high and that of oxygen is low, and there is near-total or total darkness.

Troglodytic (cave-dwelling) beetles are not found in regions where most of the more extensive caves were covered by ice sheets for long periods during the Pleistocene epoch. The one notable exception is the fungus beetle of the family Leiodidae, *Glacicavicola bathysciodes*, a troglodytic species that lives in icy lava tubes and caves in Montana. Its closest relatives are known to live in similar conditions in eastern Europe. Adult *Glacicavicola* are typical of troglodytic beetles, possessing several morphological peculiarities, such as the loss of eyes, pigmentation, and wings. They also have long legs that enable them to search large areas for food. Many cave dwellers have incredibly thin and permeable exoskeletons that readily absorb moisture from the humid atmosphere. These unique animals quickly die from desiccation if placed in normal outside air.

Lacking photosynthesis, troglodytic beetles must rely on organic materials transported into the cave system. Beetles that live near the entrances of caves often depend on bats to supply food in the form of guano (excrement) and carcasses. For beetles that live deep within caves food acquisition is an even more challenging task. Water carries small quantities of organic materials deep into caves, where they accumulate slowly over time. Troglodytic species dependent on such resources must be adapted for making do with these meager resources. Dry caves, lacking any sort of mechanism for the distribution of organics to the deepest recesses, are generally devoid of beetles.

UPTIGHT AND OUT OF SIGHT — WITHIN the macrohabitats we have already mentioned exist countless unexplored *micro*habitats, many of which lie directly under our feet. The nooks and crannies between sand grains at the seashore and in adjacent dunes, for example, support many species of beetles that undergo their life cycles entirely beneath the surface. From a beetle's perspective, these microhabitats offer ample food and shelter for themselves and their young. Several of these subterranean beetle families feed on the decaying plant material that litters any forest floor, helping to accelerate its decomposition into the basic components of organic decay to facilitate its return to the nutrient cycle. Still other families are predators, such as histerid and staphylinid beetles, whose diminutive stature enable them to seek prey easily in very small spaces.

Other microhabitats exist above ground. They, too, are discretely tucked away from the view of all but the most persistent and specialized of predators and parasites. Leaf-mining buprestids, chrysomelids, and weevils develop within the compressed environment between the upper and lower epidermal layers of plant leaves. Generally, a single egg is laid by the female either on the leaf surface or in a prepared entrance to the leaf interior. The emerging larva develops by eating some or all of the tissue of the leaf. The beetle may pupate within the leaf after the leaf has fallen to the ground, with the adult set to emerge the following spring. The entire life cycle can usually be observed in the ever-widening feeding track of the rapidly growing larva, which terminates in a distinctly expanded pupal chamber outfitted with an escape hatched nibbled away by the newly emergent adult. Leaf miners are usually host-specific, adapted to living amid defensive chemicals concocted by the plant to discourage herbivores.

{GO FORTH AND DIVERSIFY}

ONCE beetles have colonized these various habitats, their success, in terms of species diversity, is promoted further by factors that isolate populations, resulting in the evolution of novel forms. **Geographic isolation**, a result of physical barriers such as mountains, valleys, deserts, and oceans, in itself does not lead to speciation (the development of new species), but it does reduce or prevent migration and dispersal, thereby limiting opportunities to mate and restricting gene flow. Species of beetles that are geographically isolated are said to be **allopatric**. **Sympatric** species of beetles occupy the same geographically confined community but may be isolated by ecological factors, with species occupying different habitats or separated by discrete activity periods and/or behaviors. For example, differences in food acquisition habits may keep closely related species of beetles apart. Within the large buprestid genus *Agrilus* are four sympatric species found within forests of northeastern North America, each with its own distinctive niche. One species, the bronze birch borer, *Agrilus anxius*, utilizes species of birch. *Agrilus pensus*, the alder-birch borer, uses either alder or birch. The bronze poplar borer, *Agrilus liragus*, utilizes several species of aspen, while the aspen root girdler, *Agrilus horni*, uses aspen and poplar. All four species are able to exist sympatrically because they have subdivided the available resources.

Reproductive isolation may be the result of a combination of external geographic and/or ecological factors, or by intrinsic factors such as incompatible male and female genitalia. Lack of mate recognition, the inability of species to recognize each other's chemical and visual messages, also prevents couplings between members of different species.

{COMPLETING THE PUZZLE}

TO really appreciate any beetle, we must not only take into account its present distribution and environmental preferences; we must also consider the habits of its ancestors. The study of fossil beetles provides a conceptual framework for the elucidation of present patterns of distribution and dispersal and presents an opportunity to answer basic questions about rates of speciation and extinction. Reliably identified fossils from the

Quaternary period (the last 1.6 million years) have shown that some modern species of beetles are more than a million years old. Fossil specimens from the Miocene epoch (5.7 million years before present) indicate that our modern genera have been around for several million years, represented by species that may be considered the immediate descendants of modern beetle fauna. These intriguing bits of information may hold clues to our own evolution—how far have we come and how much time we have left.

The science of paleoclimatology has been advanced through the study of fossil beetles. [PLATE 49] Fossil beetles may be better than woody plants at being indicators of climatic change. The relatively short lives and the mobility of beetles allow them to react more rapidly to environmental change and keep the conditions in which they live more or less constant, despite abrupt climatic changes. Paleoentomologists studying ancestral beetles in their ecological and biological contexts have begun to appreciate the selective forces that have driven the evolution of our modern beetle fauna. If nothing else, the fossil evidence demonstrates that while vertebrates have suffered massive extinctions, presumably through climatic fluctuations, beetles continue to survive and diversify.

We must continue to forge ahead with detailed biological and distributional accounts of beetles living throughout the world, from intricate expanses of tropical rain forests to the smallest isolated patches of soil, for this data will lead to a greater understanding of our world, both past and present. Endemic, relict, or any other population of beetles with limited or restricted distributions are particularly worthy of our attention and should be conserved as invaluable scientific resources, for their evolutionary histories serve simultaneously as windows to the past and the future.

[PLATE 49]

The science of paleoclimatology has been advanced through the study of fossil beetles. Fossil beetles may be better than woody plants as indicators of climatic change. The relatively short lives of beetles, coupled with their mobility, the allow them to react more rapidly to environmental change and keep the conditions in which they live more or less constant, despite abrupt climatic changes. Fossil of a buprestid, genus *Lampetis. Photograph courtesy of Stephan Schaal, Naturmuseum Senckengerg.*

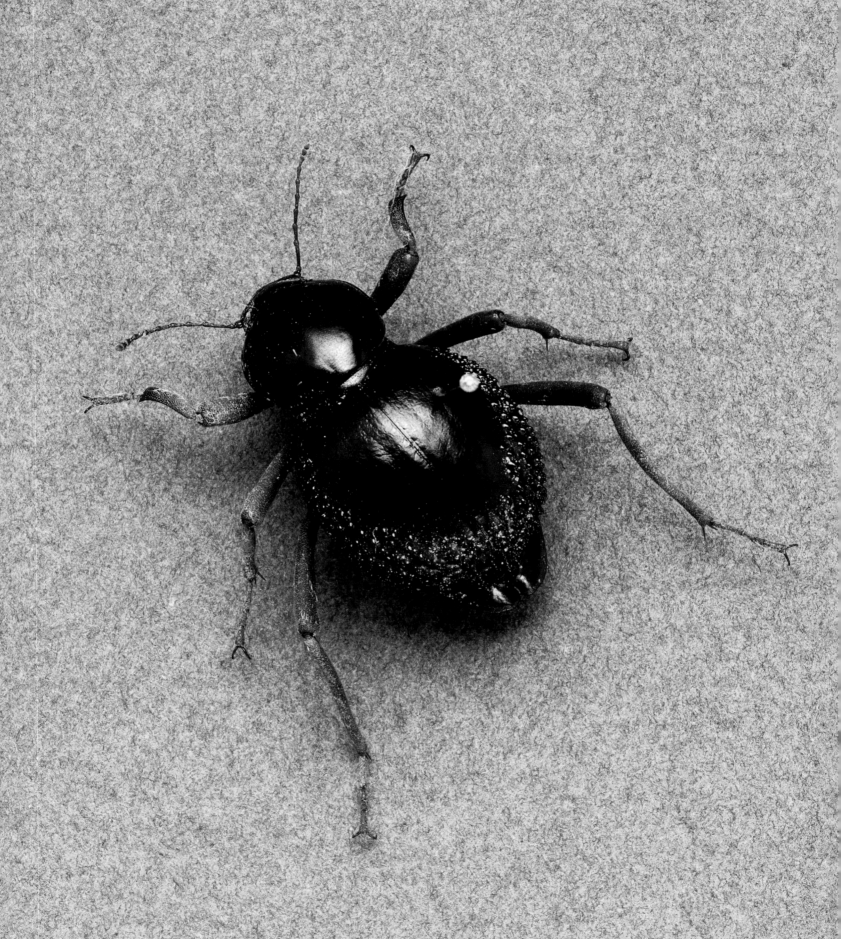

four

IT'S A BEETLES LIFE

*A male firefly blazes his trail through the woods. At last he perceives
a dim inconspicuous gleam, a mere spark, but it is his landing bea-
con and he levels off and steers straight for the wingless mate, who
has laboriously climbed to the top of a fern and there hung out her
signal: "Come oh come, so that the race of fireflies may go on!"*

William Beebe, *High Jungle*

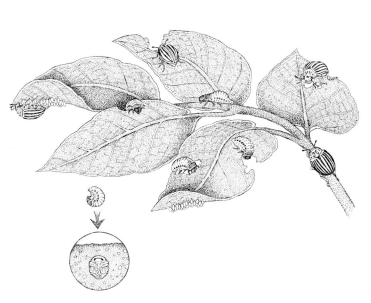

THE overwhelming number of beetle species is equally matched by the myriad ways in which they go about making a living. The life cycles and styles of beetles resemble a dance, with the dance card of each egg, larva, pupa, and adult a blur of interactions with a multitude of biotic and abiotic factors. In any single book, we can only begin to offer general remarks about the incredible diversity of ways in which beetles develop, reproduce, communicate, cooperate to raise their young, feed, defend themselves, and interact with other organisms. We know little about these fascinating creatures because they undertake only a few of their activities within our dimension of observation and understanding.

{LIFE CYCLES OF THE DIVERSE AND CHITINOUS}

THE basic development of beetles is similar to that of butterflies and other **endopterygotes**, insects whose wings develop within the larvae and that undergo complete metamorphosis. [ILLUSTRATION L] The stages of **complete metamorphosis**—egg, larva, pupa, and adult—adapt each beetle species to a specific, yet dynamic suite of seasonal and ecological conditions. Most beetles have a fairly regular life cycle, with one or more generations per year, that is in concert with seasonal changes, particularly in temperate climates and in high elevations.

Beetles produce eggs either one at a time or by the hundreds, scattering them about haphazardly or carefully depositing them on or near suitable larval foodstuffs via a long, membranous ovipositor festooned with sensory hairs at its tip. Ground-dwelling species may simply leave their eggs in the soil, placing them in dung piles or scattering them about in compost heaps.

[ILLUSTRATION L]

Most beetles, such as the Colorado Potato beetle, *Leptinotarsa decemlineata* from North America, undergo complete metamorphosis with four distinct life stages-egg, larva, pupa, and adult.

[PLATE 50]

Talking abdomens-South African tenebrionid beetles of the genus *Psammodes*, popularly known as tok-tokkies, drum their abdomens against the soil to attract mates. *Specimen courtesy by Maxilla and Mandible.*

95

Leaf grazers and leaf miners drop their eggs at the base of a plant, glue them to a stem or leaf, or insert them in a crevice in the bark. Some leaf miners exhibit the most highly evolved egg-laying behavior: The female tears the tissue of the leaf surface and deposits the egg inside. Female wood-boring beetles girdle the terminal portion of a branch with their mandibles to create new sites for tunneling larvae to develop after hatching from the eggs. In some staphylinids, chrysomelids, tenebrionids, and carabids the eggs hatch within the reproductive organs of the mother. [ILLUS-TRATION M] In this situation, called **ovoviviparity**, eggs are retained in the reproductive tract of the female until they hatch; then the larvae are "born."

Upon hatching, the first stage of the larva begins its life with a single purpose: to eat. [ILLUSTRATION N] Because they have continuous and voracious appetites, beetle larvae grow rapidly, proceeding through a series of stages, each one called an **instar**. A beetle larva may scavenge on carrion, feed within a ball of dung that was rolled by its parents, attack roots, mine the tissue between the upper and lower layers of a leaf, or chew its way through dead wood. Some wood-boring larvae take several years to complete their development into adult beetles. The larvae of bostrichids have been known to survive the treatments and processing of various woods used in furniture and house construction, slowly continuing their development and taking up to thirty-five years to emerge as adults.

Each instar is punctuated by a **molt**, the entire shedding of the exoskeleton to accommodate the insect's growth. The larva molts at least three times before reaching the pupa stage. Often inaccurately referred to as the "resting stage," the pupa serves as the vessel for dramatic biochemical transformations, which reconstruct the tissues of a nonreproductive eating machine (the larva) into a precision breeding instrument (the adult) that may or may not continue to feed. In temperate climates, the pupa—carefully tucked away in soil, humus, or within the tissues of plants—is often the best-equipped insect stage to carry the individual beetle through harsh winter conditions.

Despite the apparent universality of these four distinct stages of the life cycle—egg, larva, pupa, adult—not all beetles follow this path. Micromalthids, phengodids, [ILLUSTRATIONS O AND P] and some lycids and lampyrids exhibit **paedogenesis**, the retention of immature features. In these families the pupal stage in females has been eliminated, and the adult can be distinguished from the larva only by the presence of reproductive organs. In other families of beetles, adult females may lack hind wings and occasionally elytra, but they have all the other typically adult features, such as legs, antennae, and mouthparts. [PLATE 50]

The onset of spring or summer and the right combination of temperature and moisture from increased sunlight and rain signal the adult beetle to emerge from the pupa, ready to find a mate to continue the cycle of life. But not all species of beetles require a partner to fulfill the biological imperative of reproduction. **Parthenogenesis**, development from an unfertilized egg, is common among many species of orthopteroid insects, such as walking sticks. This mode of reproduction is rare among beetles, occurring in families such as the curculionids, chrysomelids, ptiliids, ciids, and bothriderids. Males of parthenogenetic species are rare or unknown; the females are solely responsible for maintaining the population by simply cloning themselves.

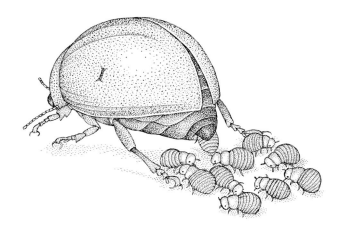

[ILLUSTRATION M]

The chrysomelid genus *Eugonycha* exemplifies ovoviviparity, where the eggs hatch within the reproductive tract of the female and then the larvae are "born".

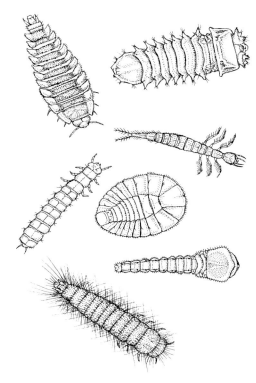

[ILLUSTRATION N]

The diversity of the Coleoptera is also reflected in the myriad of their larval forms. Upon hatching from the egg, the life of the larva is driven by one simple task, to eat. Left column, top to bottom: *Silpha, Cucujus, Dermestes.* Right column: lepturine cerambyid, Hydaticus, Psephenus, Chrysobthris.

[PLATE 51]

The adult female of the North American elaterid, *Euthysanius lautus*, lacks wings and has greatly reduced elytra, but has all other typically adult features, such as legs, antennae, and mouthparts. *Photograph by Rosser Garrison.*

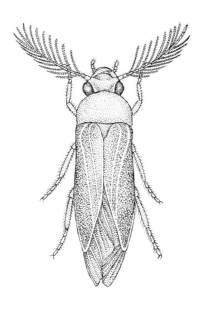

[ILLUSTRATION O]

The increased surface area of the pectinate antennae of the male *Zarhipis integripennis* are used to detect minute traces of sexual pheromones released by the female.

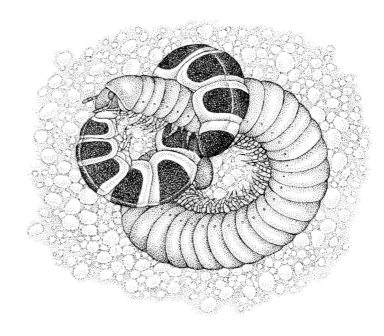

[ILLUSTRATION P]

The adult female of the phengodid *Zarhipis integripennis*, shown here feeding on a millipede, exhibits paedogenesis, the retention of the immature features of the larvae.

THE process of producing a fertilized egg in beetles is not much different from that of vertebrates: The male deposits sperm cells inside the female. Unlike vertebrates, however, relatively short-lived female beetles have an enormous reserve of eggs awaiting fertilization. The female generally needs to mate only once, although she may be courted and grasped by a number of enthusiastic males, who are often attracted en masse to the odor plume from the pheromones released by a single female. The female needs to receive enough sperm to fertilize her entire complement of eggs. To gather enough sperm, she may have an internal reservoir, or **spermatheca**, within which to store the sperm. The eggs do not become fertilized until they travel past the spermatheca on their way to being laid in a site selected by the female.

TALK TO ME LIKE all other animals, beetles communicate primarily to find mates. Whereas some species wallow about in mate-rich habitats, such as stored grains, others must disperse over wide distances, searching through tangled vegetation or layers of debris to find a mate. Beetles use various communication strategies—including mechanical, visual, and chemical tactics—to locate mates.

Several families of beetles use sound to locate one another. Passalids, cerambycids, and scolytine curculionids (bark beetles) create a shrill noise by rubbing various parts of their bodies together. This behavior is called **stridulation**. Male death watch beetles (anobiids) bang their heads against the walls of their wooden galleries to lure females. South African toktokkies, tenebrionids of the genus *Psammodes*, march to a different tune: They drum their abdomens against the soil to attract mates. [PLATE 51]

FLASH SESSIONS FLASHING lights are one of the surest ways of attracting the attention of most animals, as any casino operator in Las Vegas will attest. For centuries humans have been fascinated by light-producing, or **bioluminescent**, organisms. Among insects, beetles offer the best-known examples of bioluminescence. The delicately built, soft-bodied fireflies or lightning bugs (which are neither flies nor bugs) belong to one of three bioluminescent beetle families, the Lampyridae. [PLATE 52] The other two are the Phengodidae and Elateridae. Even the eggs and larvae of some lampyrids glow.

In 1885 the French physiologist Raphael Dubois dissected the thoracic light-producing organs of the elaterid *Pyrophorus* and discovered that two substances are involved in an enzyme-driven system that produces the light. [PLATE 53] He named these substances *luciferin* and *luciferinase* after the angel of light, Lucifer. The details of bioluminescence in beetles are complex and not entirely understood. This enzyme system is powered by adenosine triphosphate, or **ATP**. An essential ingredient of all life forms, ATP functions as the biological transporter of chemical energy, providing the extra boost all cells need to perform their various tasks. Within the light-producing cells of the firefly, the enzyme luciferinase attaches luciferin to ATP. This process energizes luciferin and allows oxygen that is pumped in through the trachea to attach to one of the carbon atoms of luciferin, kicking out an electron into a higher orbit. The luciferin releases the oxygen and the carbon as carbon dioxide. As the electron drops back into its normal

[PLATE 52]

The delicately built, soft-bodied fireflies, or lightning bugs, such as members of the somewhat cockroachlike genus *Apisoma*, belong to one of three families of bioluminescent beetles, the Lampyridae. *Specimens courtesy of the Natural History Museum of Los Angeles County.*

orbit, a bit of energy is released as a tiny flash of light. Multiply this activity by the thousands of cells found within the light-producing organ of a lampyrid and you have a flash of light that is easily visible to the human eye. This process—the use of ATP and oxygen to produce light—is the exact opposite of the process by which plants capture the energy of the sun, photosynthesis.

This form of bioluminescence is nearly 100 percent efficient; almost all the energy that goes into the system is given off as light. An incandescent light bulb, by comparison, is only about 10 percent efficient; the remainder of the energy is lost as heat. The light-producing organs of one firefly have been said to produce 1/80,000 of the heat produced by a candle flame of comparable brightness. The Roman scholar Pliny surmised that fireflies control their flashes of light by opening and closing their wings. We know now that adult fireflies control their flash patterns by regulating the oxygen supply to their light organs. The result is a species-specific pattern of light emissions.

Since ATP powers many biological reactions, systems sensitive to ATP levels can be a useful biochemical tool. NASA is using the bioluminescence system, among other things, to determine if there are life forms on other planets. A mixture of luciferin and luciferinase is included aboard exploratory spacecraft. Upon landing, a mechanical arm scoops up extraterrestrial soil and places it in the container holding this mixture. The presence of ATP in the soil, implying possible life forms, would activate the system and send a signal back to Earth.

Bioluminescence in lampyrids is generally used to signal a potential mate; each species has its own pattern and method of presentation. The color of the light usually varies from light green to orange, its intensity and hue dependent on temperature and other environmental conditions. Typically, males fly about the habitat emitting a species-specific flashing pattern, while the females remain stationary, perched on branch tips, rocks, or other promontories. When responding to a male, the female at first hesitates and then returns the message with a pulse that lasts for a predetermined time period. The number of flashes, rate, and duration of the male signal are critical to mate recognition, as is the delay and length of the female's response. The male continues to repeat his signal at regular intervals until he receives an answer from a receptive female. When he receives an answer, the male continues his flashing while flying toward the female's signal. This "conversation" continues for five to ten exchanges until the male reaches the female. The signals of *Photinus* fireflies vary from rapid, rhythmic bursts of light to long, sustained flashes over a period of time. The females of some firefly species glow continuously or respond only to continuously glowing males.

In eastern North America, two genera of fireflies are common, *Photuris* and *Photinus*. Using electronic devices to mimic the flashing patterns of these beetles, scientists discovered that two or more species that are sympatric (sharing the same habitat) have flash patterns that differ significantly, whereas the patterns of species that do not live in close proximity may be quite similar to one another. These observations demonstrate that differences in flashing patterns are designed to prevent the insects from making reproductive mistakes.

Although sympatric species occupy the same habitat, they tend to inhabit different niches. Females of one species generally exhibit a preference for a specific niche or orientation to a mating site, while males exhibit behaviors that correspond to this selectivity. This coordinated selectivity helps eliminate the confusion that may result from another species calling nearby. Restricted periods of activity, seasonally or temporally, also serve as isolating mechanisms to ensure that individuals of the same species can locate one another as easily as possible.

One of the most unusual flashing patterns is the synchronicity that has been observed in parts of tropical Asia, resulting in what is known as *firefly trees*. The trees are occupied by both males and females, but only the males participate in synchronous flashing. Occasionally a complex of different species will occupy the tree, resulting in a combination of flash patterns presumably effective for each species. The beetles maintain this behavior hour after hour, night after night, for months. The synchronicity of the flashing is the result of each beetle matching the flash patterns of its neighbors. Congregations like these are believed to increase the chances for a rendezvous between males and females, since they can be seen easily by both sexes.

There is a twist, however, to flashing as a system of mate recognition. Females of at least twelve species of *Photuris* fireflies prey on male fireflies of the genera *Photuris*, *Photinus*, *Pyractoma*, and *Robopus*. The females lure the males in by mimicking the mating signals of the prey species and then seize and eat the males. Field experiments with a penlight simulating the prey species' signals revealed that female *Photuris versicolor* is capable of returning a variety of appropriate responses.

It has been suggested that the evolution of this scheme of predatory deception can be explained by comparison of existing flash patterns of predator and prey. The flashed responses of male prey appear to be similar in delay timing to the predator's own mating responses. False signals could easily be derived originally from the mating responses and then modified. The locomotion flashes of some species—that is, the flashes emitted by both males and females when walking, landing, or taking flight—are very similar to that of the predaceous females and would require little modification.

BEING LED ASTRAY DESPITE millions of years of fine-tuning, the visual and tactile cues developed by some beetles have not adequately prepared them for some interactions with human technology. The awkward and clearly unrewarding encounters of the buprestid *Julidomorpha bakewelli* with a beverage container are a prime example. An Australian beer manufacturer unwittingly designed a bottle that has the essential appearance and feel of a female *Julidomorpha*, at least by the reckoning of the males. Since the males seem to prefer the largest females as partners, the bottles always win the contest. Concerned by the frequency of fruitless overtures of the males that were induced by the bottle, the company has redesigned the vessel to encourage *Julidomorpha* males to pursue close encounters with their own kind.

SNIFFING OUT THE ONE of the most intriguing systems for bringing
CONJUGATION the sexes together is the utilization of plumes of messenger molecules, or **pheromones**. Pheromone molecules are released into the wind, where they expand by diffusion and disperse by air turbulence. Male beetles that use pheromone systems

102

to locate females are often capable of detecting just a few molecules in the air, enabling them to find its source, a receptive female, at considerable distances, tucked away among the tangled masses of vegetation or buried beneath the flowing sands of desert dunes. Investigators have used their knowledge of beetle pheromones to design traps to detect the presence of stored grain pests or to lure them away from each other in an attempt to interrupt their reproductive processes. Males and females use pheromones to attract the opposite sex or to form aggregations [PLATE 54] in which they can display courtship behavior to increase their chances of finding a mate.

After finding each other, the male and the female copulate. Precopulatory behaviors are not common among the Coleoptera. The male usually mounts the female from above and behind to insert a marvelously sculptured organ. [PLATE 55] The genital apparatus of the males is often so distinctive that it is the key to their identity. The morphology of genitals is routinely used in identifying fossil beetles at the species level. The importance of using genital characters arose from the belief that these intricate structures must have some purpose, possibly to correspond uniquely with the apparatus of the female to the mechanical exclusion of all other species. Although this "lock-and-key" hypothesis was proposed more than a century and a half ago, its validity is still open to question today (see also "Reproduction" in Chapter 2).

Critical studies to determine whether the genitals of males and females are functionally related are rare or involve matings between species of beetles that would not encounter each other under normal circumstances. However, research on the functional morphology of copulatory organs in other groups of insects strongly suggests that there may be no correlation at all between the male and female copulatory organs; thus the "lock-and-key" hypothesis may lose favor with systematists and evolutionary biologists. Nevertheless, coleopterists still rely heavily on the unique genital characters of the male and, to a lesser extent, the female, for species identification.

Males help enhance the reproductive success of individual beetles by inseminating the partners of other males or by preventing their own partners from acquiring sperm from other males. The latter behavior is called **postinsemination association**. Beetles may remain in a copulatory embrace long after sufficient time for reproductive materials to be transferred has passed. [PLATE 56] The male tiger beetle, for example, may disengage his genitalia but continue to grasp the female with his mandibles.

{ FAMILY VALUES }

WHEN we think of cooperative family groups among the insects, what comes immediately to mind are ants, bees, wasps, and termites, the truly social insects. However, several orders of insects, including beetles, exhibit varying degrees of presocial behavior, which includes some but not all of the following characteristics: brood care, [ILLUSTRATION Q] cooperative care of young, reproductive division of labor, and overlapping generations living together and contributing to the overall well-being of the colony. Several factors have been suggested as being conducive to these cooperative behaviors: stable or structured environments, physically stressful environments, rich and short-lived resources, and predation. The payoff

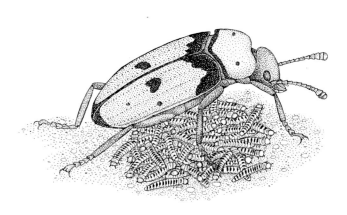

[ILLUSTRATION Q]

Ants, bees, wasps and termites are not the only insects which participate in cooperative family groups. Several families of beetles exhibit varying degrees of cooperative behavior. Females of *Pselaphicus giganteus* will protect their larvae by shielding them with her body.

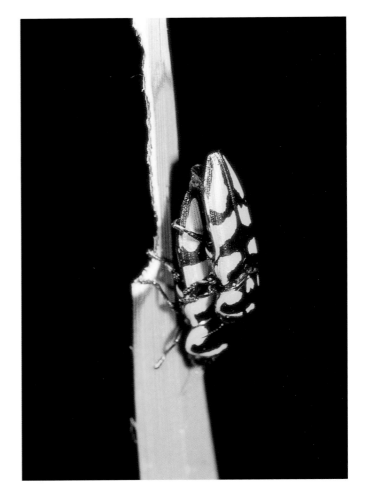

[PLATE 54, *left*]

An aggregation of cetoniine scarabs feeding on the sap of a tree in Malawi. Most of the aggregation consists of two species: *Plaesiorhinella trivittata* and *Plaesiorhinella plana*. *Photograph by Charles Bellamy.*

[PLATE 55, *right*]

A mating pair of jewel beetles from arid western Texas, *Thrincopyge alacris*, on the foliage of its host plant, *Dasilyrion wheeleri. Photograph by Charles Bellamy.*

[PLATE 56]

Beetles, such as this mating pair of melolonthine scarabs from
California, *Ceononycha parvula*, may remain in a copulatory
embrace long after sufficient time for reproductive materials to
be transferred has passed. *Photograph by Arthur Evans.*

for cooperation among social species is increased survivorship of the offspring.

Subsocial behavior—the parental care of eggs and larvae—is known in at least ten families of beetles. It is rarest among detritivores (which feed on debris), more common among predators, and best known in beetles that feed on carrion or dung. The female may carry the eggs beneath her body until they hatch, or she may place the eggs in a specially constructed chamber and defend it. One of the most simplistic examples of subsociality in beetles is demonstrated by a few female carabids, which construct depressions in the soil where they deposit their eggs to protect them from predators and clean the eggs to prevent their loss from damage by fungi until they hatch.

Brood care has been observed in several chrysomelid species, [PLATE 57] including the neotropical cassidine chrysomelid *Acromis sparsa*. The female of *Acromis* carefully selects a young leaf tip of its host plant and weakens the leaf slightly by chewing the midrib. She then attaches a cylindrical clutch of eggs under the protection of the drooping leaf and remains with them until they hatch. The newly hatched larvae remain in a group to feed, always with the mother in attendance. When the larvae stop feeding, they group together so that the mother can climb on top of them to shield them from predators and parasites with her body. Inevitably, the feeding activities of the larvae deplete the leaf, forcing them to find other sources of nourishment. During these searches the larvae are particularly vulnerable to attack, but the mother is never far behind these marches and soon catches up to the group as they settle down to feed on another leaf. The mother continues to remain with her offspring through pupation and into adulthood. Although there are advantages to this strategy, *Acromis* offspring are still susceptible to egg parasites and larva predators.

Passalid beetles have long been thought to exhibit subsocial behavior, although much of their interactions remain a mystery. All developmental stages of passalids may be found tunneling within rotting wood, interacting with one another. The larvae of the eastern North American *Odontotaenius disjunctus* consume decaying wood but seem to fare better ingesting wood that has been chewed, predigested, or converted into frass by adult beetles. Adults and larvae living in dense colonies stridulate continuously, leading some coleopterists to speculate that this form of communication helps keep groups of adults and larvae together. Seventeen different types of sounds have been described in the adults and larvae, representing what is possibly the most complex sound communication system known in arthropods.

Other beetles exhibit different levels of parental care. Female staphylinids of the European genus *Bledius* maintain and defend their intertidal brood tunnels, providing the larvae with algae for food. Bark beetles, now included in the family Curculionidae, and some scolytids carve elaborate galleries, sometimes with a central chamber connecting individual "cradles," or brood chambers, in the trunks of trees and cultivate a symbiotic fungus for food, much as some ants and termites do. Interestingly, the fungi are beetle-specific, introduced to the tree solely by the beetles, and serve as food for both adults and larvae. Adult beetles possess mycangia, specialized pits for storing fungal spores as they move about the forest trees. Although these beetles may be communal, no form of communication between adults and larvae is evident.

[PLATE 57]

Cassidine chrysomelids, such as this unidentified pair of African tortoise beetles, are characterized by the nearly circular outline of their body, the edges of which are flared out and flattened. When disturbed these beetles tuck in their head, legs, and antennae and tightly appress themselves to the smooth surface of a leaf. Several species of cassidines exhibit brood care. *Photograph by Charles Bellamy.*

Examples of advanced subsociality in beetles are found among the dung-feeding and carrion-feeding families Staphylinidae, Geotrupidae, Scarabaeidae, and Silphidae, whose behaviors include cleaning of the nest, progressive feeding of the young, and even division of labor among the adults. In the last three families, males and females cooperate in digging nests for their eggs and supplying them with dung or carrion. Dung and carrion provide a nutrient-rich medium for adult beetles to rear their brood. These resources, however, are widely dispersed and short-lived. The competition for dung and carrion among beetles, flies, ants, and other scavengers is intense and has resulted in the evolution of behaviors designed to sequester this valuable resource. The "spoils" are typically buried in a subterranean chamber to exclude most predators and parasites and to maintain optimum moisture levels for successful brood development.

Some of the best-known nest builders are among the scarabaeine and aphodiine dung scarabs. Since ancient times these seemingly industrious creatures have been observed carefully cutting, shaping, rolling, and burying pieces of animal excrement. [PLATE 58] Meticulous observations have revealed three primary strategies employed by dung beetles to secure and sequester dung for their young. The simplest strategy is simply to live and breed directly in the dung pad, without any special preparation or nest building. This method is employed by all aphodiines and some scarabaeines, which are thus collectively referred to as *endocoprids*. *Paracoprids* carve out chunks of dung and stuff them into burrows underneath or immediately adjacent to the dung pad. *Telocoprids* quickly carve out pieces of dung and shape them into balls that can be easily rolled away. Some paracoprids and telocoprids dig elaborate burrows that they then stuff with plugs of dung or carefully supply with delicately formed dung balls that serve as food either for larvae or for adults.

Beetles of the silphid genus *Nicrophorus* exhibit the most advanced behaviors of subsociality known in the Coleoptera. [ILLUSTRATION R] Not only do adult pairs of *Nicrophorus* construct nests for laying their eggs; they continue to supply food to the larvae after they hatch. Either the male or the female initiates the carrion burying process, during which time the arrival of a mate is quite likely. If the carcass is encountered on soil that is too hard for burial, the beetles randomly search for a more suitable burial site and may move it as much as several meters by lying on their backs underneath the carcass and using their legs as levers. The beetles mate only after the carrion has been secured in a chamber beneath the ground. The carcass, such as a mouse or bird, is meticulously prepared for consumption by the larvae by being rendered into ball after having its skin and appendages removed. The carcass serves as food for both adults and larvae. The female firmly packs the floor and walls of the subterranean chamber, and she lays her eggs along the walls of a vertical chamber dug directly above the carcass. A conical depression created on the upper surface of the carcass serves as a receptacle for droplets of regurgitated tissue deposited by both the male and female. The resultant elixir accumulates and eventually serves as food for the newly hatched larvae.

Adult *Nicrophorus* are known to stridulate during stress, copulation, confrontations with other beetles, and to communicate with the larvae. The female calls the larvae to the pool of regurgitated tissue with a sound clearly audible to the human ear by rubbing a ridge on the elytron against the

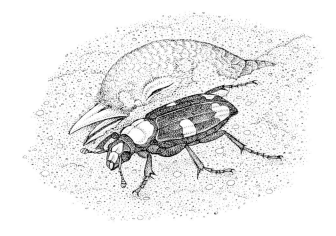

[ILLUSTRATION R]

Beetles of the silphid genus *Nicrophorus* exhibit the most advanced cooperative behavior known in beetles. Either the male or female will initiate the preparation of a carcass for burial, during which time the arrival of a mate is quite likely. The pair will construct a nest for egg laying and will feed the larvae after they hatch.

[PLATE 58]

Dung rolling scarabs carefully cut and shape animal excrement into balls which serve not only as a brood chamber for a single larva, but as its food source as well. This brood ball was formed by the genus *Kheper* in South Africa.
Photograph by Arthur Evans.

[PLATE 59 *left*]

One of the clearest reasons for the success of beetles is their ability to use a wide variety of resources as food, such as fungus. Many species of the family Erotylidae, the fungus beetles, are often brightly colored with red or yellow spots with zigzagging colored lines. Adults and larvae live and feed on fungi on the ground or growing up on the bark of trees. Fungus beetles may exude a foul-tasting fluid and/or feign death to avoid predation. *Specimens courtesy of the Natural History Museum of Los Angeles County.*

[PLATE 60 *right*]

Beetles consume the full range of plant tissues and structures, living and decaying, from canopy foliage and flowers high above the rain forest floor to subterranean roots wedged between slabs of solid stone. The mole beetle, *Hybocephalus armatus*, is an aberrant member of the family Cerambycidae that is restricted to the state of Bahia in Brazil. It is preseumed that the larvae of this beetle are root feeders, but the fossorial nature of the adult males remains a mystery. At one time, local women were known to have tied them to a ribbon and attached them to cradles for the amusement of infants. *Specimen courtesy of Maxilla and Mandible.*

[PLATE 61]

The small and very rare buprestid, *Hesperorhipis mirabilis*, from the Colorado Desert of southern California, at rest on a thorn of its host plant, mesquite. *Photograph by Charles Bellamy.*

corresponding abdominal segment. Studies have shown that young larvae will orient themselves to a recording of the adult's stridulation. The larvae receive parental care throughout their development, and only upon pupation do the adults voluntarily leave the brood chamber.

{THE BEETLE BUFFET}

ONE of the clearest reasons for the success of beetles is their ability, as a group, to use as food a wide variety of resources in nearly all habitats. [PLATE 59] They function as herbivores, predators, parasites, and scavengers, yet most species feed strictly on plants and often on only one species, genus, or family of plants.

Life on the flowering plants may have presented a formidable evolutionary hurdle, but once cleared, the interface between plants and herbivores provided the template for the dramatic radiation of all terrestrial animal diversity. Russian paleoentomologists have inferred that insects, as a group, began as herbivores, feeding initially on the reproductive organs or spores of nonflowering plants or fungi present during the Carboniferous period; there is no fossil evidence of leaf feeding by insects during this period. Beetles first appeared in the geological record in the Lower Permian period (2350 million years before present), and some scientists believe that they fed on fungi under bark or in rotting wood. Signs of feeding damage on leaves—probably by worms, snails, or millipedes as they consumed fallen leaves—and wood boring first appeared in Permian plant fossils and became more common in the Triassic period. By the mid-Triassic, the main evolutionary lineages of modern beetles were established.

The opportunities for herbivory increased with the diversification of angiosperms in the Cretaceous period (125 million years before present). At this time most herbivorous beetles had larvae that lived inside of and fed on their plant host, where they were protected from predators and parasites. Today, beetles use and consume the full range of plant tissues and structures, living or decaying, from subterranean roots wedged between slabs of solid granite to the canopy foliage and flowers of the tallest rain forest trees. [PLATE 60]

A constant war rages between plants and herbivorous insects, with plants often using potent chemical or physical deterrents to repel insects. Some beetles may circumvent chemical deterrents by chewing open wounds on leaves or by girdling stems to allow the toxic sap to drain away before they feed or lay their eggs. Other species have developed the ability to sequester toxic chemicals in their own bodies, incorporating the plant's chemical defenses as their own protection.

A BORING EXISTENCE WOOD boring as a feeding choice was perhaps the most primitive condition in prehistoric beetles. Today, the most diverse groups that tunnel in dead or dying plant tissue are the bostrichids, [ILLUSTRATION S] buprestids, [PLATE 61] cerambycids, [PLATE 62] and curculionids. A dead tree branch presents several feeding niches: stems of different sizes, and different layers of tissue from the bark inward. As a dead branch ages, a succession of different beetle species work in concert to complete the recycling process.

Some buprestids arrive within hours of the cutting of a branch. Members of the genus *Melanophila*, which have sensory structures that are sensi-

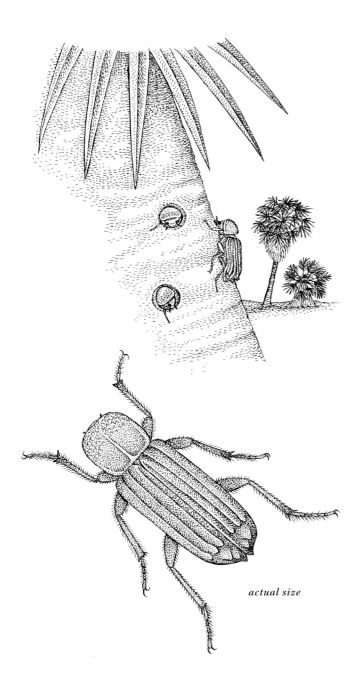

actual size

[ILLUSTRATION S]

The giant palm borer, *Dinapate wrighti*, is known from Arizona, California, and the peninsula of Baja California, Mexico. Although originally restricted to palm oases and highly prized by collectors as rare, the proliferation of palms as street trees has resulted in their spread and in some areas they are considered pests.

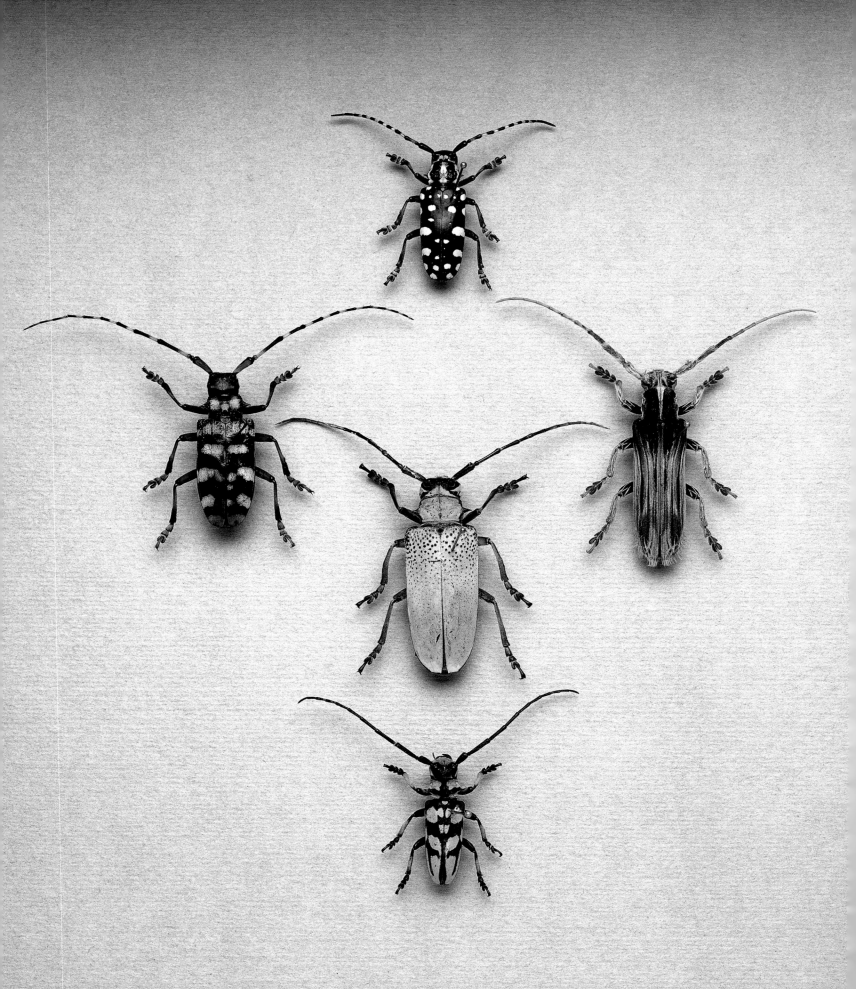

The long-horned wood boring beetles, family Cerambycidae, are usually cylindrical, elongate beetles with antennae at least two-thirds of their body length. Their long cylindrical larvae usually feed internally on the branches, trunks, or roots of shrubs and trees. Top: unidentified species; left: *Anoplophora medembachi*; center: *Calostema sulphurea*; right: *Xylorhiza adusta*; bottom: unidentified species. *Specimens courtesy of Maxilla and Mandible.*

[PLATE 63]

Dung beetles, such as the African *Garreta nitens*, use their serrated foreleg and broad hoelike head to carve out chunks of fecal material to use as food. They ingest huge quantities of dung, using their membranous mandibles to strain out remnants of undigested food, bacteria, yeasts, and molds. *Photography by Charles Bellamy.*

tive to infrared radiation, are first attracted to their hosts by the heat of fire. *Melanophila* beetles can be seen running along still-smoldering branches, wood that other beetles will not find attractive for several years.

Endosymbiotic microorganisms, such as bacteria, yeasts, and fungi, assist various wood-boring beetle groups with the digestion of the primary component of plant food, cellulose. These symbiotic organisms generally reside in outgrowths of the midgut, although in the scarabaeoid groups that develop on decaying wood, the fungal symbionts live in a fermentation chamber in the hindgut. The symbionts are transferred from mother to egg as the egg passes through a residue in the ovipositor. Upon hatching, the larvae immediately consume their own eggshell, which is laden with microorganisms that are essential for the digestive system of the young beetles.

GRAZERS One of the beetle families that are most closely associated with foliage feeding is the Chrysomelidae, or leaf beetles. Chrysomelid larvae can completely defoliate or skeletonize most of the leaves on a plant and may be considered serious agricultural pests. The plant food that has attracted the greatest diversity of beetle species, however, is the flower. In their search for energy-rich pollen and nectar, beetles have become important pollinators of some groups of plants. Beetles that feed on nectar in flowers may have specialized mouthparts especially suited for extracting their favorite food deep within the corolla (the collective name for the petals).

A group of stigmoderine buprestids from South America and Australia have evolved almost tubular mouthparts that enable them to pump nectar produced by and stored in the large cup-shaped nectaries (glands that secrete nectar) of eucalyptus trees and shrubs. The maxillae of trichiine scarabs are modified into brushes to aid in their consumption of fine pollen granules. These and many other species of flower-visiting beetles are adorned with dense hairs, or setae, which inadvertently brush through the stamens (the pollen-producing male reproductive structures of the flower) and retain pollen, which is later delivered to other flowers. Generally beetles are not pollinators in the same class as bees, but they play a significant role in pollen transfer in some plant groups that are not frequently visited by more traditional pollinators.

RECYCLERS AND SANITATION ENGINEERS Beetles are a major component of the FBI—fungi, bacteria, insects—the primary organisms that physically and metabolically break down plant and animal materials into their basic components to be recycled by other plants and animals. If not for the role played by carrion and dung beetles, we would soon be overwhelmed by filth and death. Carrion and skin beetles, which consume dead animal tissue, skin, feathers, fur, and hooves, include members of the silphids, clerids, dermestids, and trogids. Because of their ability to remove tissue from bone, dermestids are used by nearly every natural history museum around the world to clean animal skeletons for use in research collections and exhibits.

Dung beetles are one of the most beneficial and least appreciated of all insect groups. [PLATE 63] The amount of organic waste produced by vertebrates, especially large herbivores, attracts a variety of these industrious beetles ready to use this smelly stuff for their own food and, in the process, reduce the number of breeding sites for obnoxious flies. Dung scarabs use their serrated forelegs and broad hoelike head to carve out chunks of fecal material to be used as food. They ingest huge quantities of dung, using their membranous mandibles to strain out remnants of undigested food, bacteria, yeasts, and molds.

Many dung scarabs are generalists, flying considerable distances in search of any type of fecal material of the right age and consistency. A few species, however, are specialists, preferring the dung of just one animal. The Central American *Uroxys gorgon* and *Uroxys metagorgon* attach themselves to the fur of their preferred dung source, the three-toed sloth. The sloth, which feeds high up in the rain forest canopy, keeps a low profile among predators by coming down to the ground to defecate only occasionally and then quickly burying its dung. Even in the tropical climate, the pellets soon dry out, becoming useless to the beetles. By being attached directly to its food source, *Uroxys* can disembark to take full advantage of its favorite food while it is still fresh and nutritious.

A similar adaptation has arisen independently in Australia, where at least six species of the genus *Onthophagus* have evolved claws adapted for grasping the fur of wallabies and the large rat kangaroo. As many as 175 beetles have been collected from a single wallaby. Beetles have been observed to collect around the anus of their host, where they converge on an emerging dung pellet, grasp it, drop to the ground with it, and begin burying it.

Not all dung beetles, however, feed on dung. Many dung beetles are opportunists and take advantage of carrion. Others are specialists, preferring carrion, fungi, fruit, millipedes, and the slime tracks of snails over feces.

PREDATORS Most predatory beetles belong to families of the more primitive suborder Adephaga, most notably the Carabidae, which includes the predatory ground beetles and tiger beetles (formerly the family Cicindelidae). [ILLUSTRATION T] These animals are generally quick on their feet, adapted for running or climbing, using rather pronounced mandibles to capture and tear apart their prey. For their size, the larvae of tiger beetles are one of the most gruesome looking creatures imaginable. They are equipped with very large scythelike jaws. Their rather stout body is anchored inside a vertical burrow by abdominal hooks. When an insect comes within reach of the entrance to the burrow, nearly half of the body lunges outside the burrow to grab and drag the prey into the subterranean chamber, where the beetle can consume its victim in relative safety.

Most predatory beetles attack a broad range of other beetles, insects, and invertebrates, although some are specialists, such as some carabids that attack and kill only snails. Staphylinids and histerids seek their prey among the labyrinth of debris and decaying organic matter on the ground. Some, however, are incredibly specialized, living among ants, termites, and the fur of small mammals.

Even families of beetles that typically feed on plants have evolved groups that prefer flesh. Within the scarabaeoid families, the predatory habit has evolved independently in the cetoniines and dynastines. The cetoniine tribe Cremastocheilini includes several genera that live with ants, bees, and occasionally wasps, feeding on, among other things, broods of young and adults that are dead or dying. *Oplostomus* can be found feeding

[ILLUSTRATION T]

Many predatory beetles belong to the family Carabidae, which includes the tiger beetles, such as this impressive hunter, *Ambycheila cylindriformis*, from New Mexico. The larvae of this and other tiger beetles are some of the most gruesome looking creatures imaginable, equipped as they are with scythelike jaws that are used to impale insect prey which venture too close to their vertical burrows.

on the broods of paper wasps, while other African and Indian crematocheilines prefer spider eggs or scale insects. Members of the dynastine tribe Phileurini are also predators and have been observed feeding on insects. Two African genera, *Rhizoplatys* and *Macroyphonistes*, have been found in the brood chambers of domestic and wild honeybees, having bored through wax combs to locate and consume the brood.

{HOW TO AVOID BECOMING DINNER}

BEETLES themselves are prey for a variety of animals, yet each beetle life stage is protected by structural and behavioral adaptations. Adult beetles often rely on their mobility and a thick exoskeleton, along with various morphological modifications, as their first line of defense. Some species have an incredibly thick body wall combined with a somewhat compressed anatomy, which enables them to flatten themselves against whatever surface they are on, thereby making it a formidable task for small predators to capture them. The predator generally gives up the attack, allowing the beetle to retreat to safety. One such species, a zopherid commonly referred to as the ironclad beetle, *Phloeodes pustulosus*, can be encountered in abundance beneath the bark of oak trees in the southwestern United States. In addition to having an incredibly hard exoskeleton, this beetle defends itself by practicing **thanatosis;** that is, it plays dead to avoid attracting further attention to itself.

Many of the larger horned scarabs appear frightening and seem well equipped to repel an attack. Some of the larger lucanids and cerambycids have large, powerful mandibles capable of inflicting a painful bite. [PLATE 64] Most smaller beetles opt either to fool their enemies by looking like something that would not be considered edible, or by resembling another species that is known to be hard to capture or outright dangerous to experienced predators. Still other species camouflage themselves or possess an arsenal of noxious chemicals to discourage predators.

LIFE AS AN ILLUSION BEETLES employ a broad array of tactics to fool their enemies, including mimicry, camouflage, and warning coloration. In its most fundamental aspect, **mimicry** is any behavior that enables an animal to avoid predation by adopting the characteristics or appearance of another animal. The resemblance may be in terms of shape, color, pattern, or behavior. The concept of mimicry has been refuted in some circles on the grounds that we can never know how the true predators of these organisms perceive their potential victims. Some people suggest that the evolutionary process that might allow a gradual convergence of shape or color probably does not bestow much advantage to the intermediate forms, bringing into question the driving force of mimicry. Whether mimicry is a construct of nature or another example of the human psyche seeking to find something that isn't there, the many examples of beetles as both models and mimics are fascinating to behold.

From 1849 to 1860, English naturalist Henry Walter Bates explored the Brazilian forests to collect and observe insects. Bates noticed that the wings of unrelated heliconiid and pierid butterflies often shared similar patterns and colors. He postulated that there must be some advantage derived by the pierids, which affect the appearance of the slow-flying, yet conspicuous heliconiid butterflies. Bates rarely saw a bird attempt to take these butter-

[PLATE 64]

The antler-like mandibles of stag beetles, family Lucanidae, particularly the males, may be monstrously developed. Although some species are capable of inflicting a painful bite, the shorter and stronger mandibles of the females are generally more powerful. All of the following species are from Malaysia. Left to right, top to bottom: *Cladognathus giraffa*; *Hexarthrius deyrollei*; *Odontolabis femoralis*; and *Dorcus titanus.* · *Specimens courtesy of Maxilla and Mandible.*

flies and determined that there must be something distasteful or repellent about them. This form of deception, known as **Batesian mimicry**, has three components: a species, or model, that is unpalatable to predators; a second species, or mimic, that is palatable, but more closely resembles the model than its own relatives; and predators that have learned to associate the pattern of the model with unpalatability.

Batesian mimicry is the most common type of mimicry observed in beetles. The most well known mimicry complexes include distasteful models from the families Cantharidae, Lampyridae, Lycidae, and Meloidae. In the southwestern United States and adjacent Mexico, the cerambycid *Elytroleptus* not only mimics lycid beetles of the genus *Lycus;* it also eats them. In Australia, certain buprestids mimic diurnal flower-visiting cantharids, and some buprestids and belids mimic lampyrids. In Chile, a complex of beetles belonging to the families Tenebrionidae, Buprestidae, Cerambycidae, and Elateridae mimic presumably distasteful lampyrids. In the subtropics of the Western hemisphere there is a repeated mimicry pattern between the model leaf beetle subfamily Clytrinae, whose members share a specific pattern of dorsal coloration and covering, and their mimics, which include at least seven different genera of buprestids.

The cactus-feeding cerambycid *Moneilema* resembles in appearance and behavior foul-smelling and -tasting tenebrionids such as *Eleodes*. When disturbed, *Eleodes* raises the tip of its abdomen to release a defensive secretion. Tests have shown that species of *Moneilema* behave similarly when faced with predators such as lizards, wood rats, and skunks. Although these cerambycids can be found wandering about on the ground, their preferred spiny host, the cholla, affords them a considerable degree of protection.

Mimicry complexes do not necessarily develop around an unpalatable model; they may be patterned after models that are difficult to catch or that defend themselves by biting, stinging, or releasing noxious chemicals. Many cerambycids are striking mimics of ants, bees, wasps, and spiders. In addition to bearing similar colors as stinging or biting arthropods, the beetle's pattern of color often gives it a form similar to that of its model. Flower-visiting lepturine cerambycids not only have elongated, tapered, wasplike bodies; their flight behavior also mimics that of the model. The North American *Ulochaetes leoninus* looks and behaves just like a bumblebee. [PLATE 65] In fact, when these beetles are captured, they attempt to "sting" with their ovipositor. Many species of checkered beetles (Cleridae) also mimic pugnacious ants and a family of wasps known as velvet ants.

Several species of neotropical Curculionidae, Mordellidae, Anthribidae, Buprestidae, and Cleridae are thought to mimic flies. According to one theory, in this form of mimicry predators have learned that quick-moving flies, and species that resemble them, require a greater amount of energy to pursue and capture than would be gained by consuming them. Many fly mimics tend to converge on a scheme of coloration that includes a red head or anterior region to mimic the large red eyes of flies; a bicolored, striped middle body region; and a dark posterior region that suggests the dark wings of a fly.

German zoologist Fritz Müller studied butterflies throughout Brazil toward the end of the nineteenth century. He observed that two unrelated, inedible butterflies occasionally share a remarkable similarity and suggested that predators could learn to recognize the color patterns of these inedi-

[PLATE 65]

Lepturine cerambycids, such as the American *Ulochaetes leoninus,* and the Chilean *Callisphyrus macropus,* seen here with wings open and closed, not only have elongated, tapered, wasp-like bodies, their flight behavior also mimics the model. *Specimens courtesy of the Natural History Museum of Los Angeles County.*

ble species. **Müllerian mimicry** differs from Batesian mimicry in that all the species in the complex are unpalatable, they cannot be distinguished by predators, and they must be found in the same place at the same time. Like Batesian mimicry, Müllerian mimicry requires that predators associate the common pattern exhibited by all members of the complex with unpalatability.

If the inedible species were the only one with its pattern of warning colors, it would have to sacrifice many individuals to the predator before the predator learned to recognize the pattern of warning colors, and this carnage would have to be repeated for each inedible species with a different pattern. If two or more distasteful species are similar in appearance, however, the learning curve for the predator would result in the loss of fewer individuals of each of the distasteful species to predation before the predator learned to avoid all individuals of a similar appearance. Brilliant colors and distinct patterns that serve as a warning, often expressed by combination of black with reds and yellows, are referred to as **aposematic coloration**.

Beetles of the family Lycidae are probably the best-known Müllerian mimics of the Coleoptera; probably all members of this family are unpalatable. [PLATE 66] The closely related lampyrids and cantharids also sport aposematic colors, are capable of releasing noxious chemicals when roughly handled, and often serve as models for Batesian mimicry complexes that include other families of beetles and some moths.

Without careful observation and testing, it may be difficult to prove whether a particular mimic is demonstrating Batesian or Müllerian mimicry or a combination of both. A group of dung scarabs from the Namib Desert in southwestern Africa is an interesting case. One species differs from other local species within the same genus in that it has orange elytra, in contrast to the normally black elytra of its relatives. This species is a very fast and agile flyer, and it has been suggested that the orange coloration is an "advertisement" of this superior agility and speed. Two other flightless Namib dung scarabs, which usually possess black coloration in other parts of their range, have evolved localized populations with orange elytra that resemble those of their fast-flying relatives. These mimics are most likely protected from predators that have learned of the low reward for chasing the elusive, fast-flying models. This "speed mimicry" fits the criteria for both Batesian and Müllerian mimicry.

Another example of speed mimicry, and probably Batesian mimicry, is exhibited by the scarab genus *Lichnanthe*. These hairy beetles fly fast and low over the ground, resembling bees in flight. The buzzing sound of these and other beetles, created either through flight or by stridulation, may give a moment of pause to some predators that have had a painful encounter with stinging insects that produce similar sounds.

CRYPTIC AND
STARTLE COLORATION

Species that employ protective coloration, or **crypsis**, resemble tree bark, leaves, lichens, or inanimate objects such as animal feces. [PLATES 67 AND 68] Most buprestids are brilliantly colored, but one large Madagascan genus, *Polybothris*, has a dark, often mottled dorsal surface that probably camouflages it from predators. The larvae of the cassidine leaf beetles, or tortoise beetles, have a pair of posterior abdominal appendages that accumulate fecal pellets. When threatened, the larva holds these appar-

[PLATE 66]

Net winged beetles of the family Lycidae are probably the best known Müllerian mimics among the beetles. Most, if not all species of the net winged beetles are presumed to be distasteful. This unidentified species is from Malaysia.
Photograph by Charles Bellamy.

[PLATES 67 AND 68]

Cryptic beetles employ protective coloration to resemble tree bark, leaves lichens, or inanimate objects such as animal feces. The beautiful and cryptically colored South African jewel beetle, *Asymades transvaalensis*, is shown here resting on its host plant, *Boscia albitrunca*. *Anadora cupriventris* is another cryptically colored jewel beetle from southern Africa. *Photographs by Charles Bellamy.*

A recently described jewel beetle from western Mexico,
Pilotrulleum largartiguanum, is so-named because its cryptic
coloration suggests the appearance of lizard feces. *Photograph
by Charles Bellamy.*

ently distasteful structures up over its body. Other species of chrysomelids and buprestids have a mixture of irregular body surfaces and multicolored patches of setae that make them resemble the fecal material of other animals, including that of insects. A recently described buprestid from western Mexico that looks somewhat like feces was given the specific epithet *lagartiguanum*, which is Latin for "lizard feces." [PLATE 69]

Colors other than those considered to be aposematic can be used to startle would-be predators. The elaterid genus *Alaus* and the buprestid genus *Lampropepla* have **eyespots** (spots of color resembling eyes) that may allow the possessor a brief period to escape from predators that are momentarily surprised by outsize eyes looking at them. Many somber-colored buprestids, such as species of *Chrysobothris* and *Polybothris*, in lifting up their elytra to take flight, reveal flashes of bright iridescent blue, green, or red, borne by the abdominal tergites. [PLATE 70] Potential predators may be momentarily stunned by the sudden appearance of these bright colors, allowing their prey to escape.

CHEMICAL DEFENSE   Beetles generally advertise the fact that they contain noxious chemicals with aposematic coloration or with distinctive chemical-release behaviors, such as standing on their head, as some tenebrionid beetles do. Beetles use defensive chemicals for many different functions—as repellents, insecticides, or antimicrobial agents—directed against a wide range of target organisms. The chemical components are usually synthesized by complex glands and stored in special chambers or within the blood; in the latter case, the chemicals are released involuntarily by reflex bleeding. Some species take advantage of the chemical defenses of their host plant, incorporating the plant substances into their defense system by shunting these harmful compounds out of their digestive tract and into their body wall.

Ground beetles (the Carabidae) possess defensive glands at the posterior end of their body that produce a variety of hydrocarbons, aldehydes, phenols, quinones, esters, and acids, which are released as a stream from the abdominal opening. The large African carabids *Anthia* and *Thermophilium* are brimming with formic acid, the same chemical deterrent employed by and named after ants, and they deliver their caustic loads with amazing accuracy from the tip of the abdomen. Bombardier beetles have evolved separate chambers to store the components of their chemical defense system. By mixing hydroquinones, hydrogen peroxide, peroxidases, and catalases in a separate chamber, these beetles create a synergistic reaction that explodes out of the body with an audible pop, producing a small, yet potent, cloud of acrid spray that is heated to a temperature of about 100°C. This noxious cloud can be accurately aimed at the bombardier's enemies with the aid of an incredibly flexible abdominal turret.

Blister beetles produce cantharidin, which is secreted by reflex bleeding from around the membranes between segments at the base of their legs. [PLATE 71] Cantharidin, used as a drug by humans, is a powerful blistering agent, insecticide, and feeding deterrent and can cause blistering and sores when applied, internally or externally, to the softer tissues of vertebrate predators. Curiously, only the males and larvae are capable of synthesizing cantharidin; adult females acquire their protective chemicals via copulation with caustic males. Pyrochroid and anthicid beetles actively seek their can-

[PLATE 70]

Upon lifting its elytra, the sudden appearance of the bright metallic blue-green color of the abdominal sternites of the genus *Chrysochroa* is thought to momentatily stun predators and enhance the beetle's chances of escape. *Specimen courtesy of Maxilla and Mandible.*

tharidin-bearing cousins, chewing on their elytra presumably to profit from the toxic chemical acquired through this diet, although cantharidin may also have some sexual attraction for these beetles.

Many tenebrionids typically expel noxious quinones, standing on their head to facilitate delivery, which behavior inspires one of their colloquial names, the clown beetles. North American pack rats of the genus *Neotoma* are not deterred by this beetle's odiferous offerings; they simply grab the beetle with their paws, forcing the tip of the abdomen into the soil and eagerly begin to consume the tastier head and thorax.

Leaf and long-horned beetles may sequester noxious or poisonous chemicals from their host plants, such as nightshades, cucurbits, or milkweeds, and use these borrowed deterrents for their own protection. Adult and larval chrysomelids have evolved a broad array of chemical defenses and delivery systems. The larvae employ reflex bleeding, toxin-secreting spines, and noxious droplets that can be discharged and then withdrawn back into the body. The neurotoxic qualities of the blood of the alticine chrysomelid *Diamphidia* were well known to the African bushpeople, who used it to poison their arrows. In addition to having formidable chemical defenses, adult chrysomelids may protect themselves from predators by jumping behavior, cryptic coloration, sharp spines, reflex bleeding, or by simply tucking in their legs and dropping out of sight.

{BEETLES AS PARASITES AND SYMBIONTS}

THROUGHOUT their marvelously diverse history, beetles have frequently developed compulsory relationships with other organisms. Each individual beetle may be viewed as a universe unto itself, home to countless numbers of organisms inside and out. Many beetles, particularly those that consume wood, utterly depend on internal organisms, endosymbionts, to metabolize their food. Among the best-known external, or ectosymbiotic, relationships—both with fungi—are found in the families Curculionidae and Leiodidae. Some curculionid weevils participate in one of the most beautifully intricate relationships: These beetles carry gardens on their backs, which in turn provide habitat for several phyla of animal scavengers, commensals, predators, and parasites.

THE WALKING GARDENS OF PAPUA NEW GUINEA The bodies of some beetles in South America and Africa provide fertile ground for some plants, but it is in the moist, cool forests high in the mountains of Papua New Guinea that this association between beetles and plants is best known. At least twenty species of the large wingless weevils of the leptopiine genus *Gymnopholus*, as well as several species of cryptorynchine weevils, demonstrate one of the most striking examples of **epizoic** plants (plants that attach themselves to the outside of animals). [ILLUSTRATION U] The gardens may camouflage the beetles from their enemies, such as birds, but the specific predators the beetles in their mountain retreats are unknown.

The pronotum, elytra, and, rarely, legs of these weevils inhabiting the moss-forest zones are especially suited for supporting at least twelve different families of plants, the species of which occur widely and are often found elsewhere than on the weevils. The highly modified scales, setae, concavities, pits, and grooves of the weevil combine with ample quantities of shade

[ILLUSTRATION U]

The modified scales and concavities of the dorsal surfaces of some weevils, such as *Gymnopholus lichenifer* from the wet montane forests of Papua New Guinea, provide holdfasts for several families of plants, which are in turn home to nematodes, mites, and bark lice.

[PLATE 71]

Blister beetles, such as this African *Mylabris*, produce cantharidin, which is secreted by reflex bleeding from around the membranes at the base of the legs. Cantharidin is a powerful blistering agent and insecticide. *Specimen courtesy of Maxilla and Mandible.*

and moisture to create a microhabitat that is ideally suited for the growth of algae, fungi, lichens, liverworts, and mosses. These gardens change as the beetle ages, and mature beetles often appear worn, the result of having lost their specialized structures and thus the necessary surface for growth of the plants. Interestingly, species of *Gymnopholus* that live at lower elevations not only lack plants, living as they do in drier, less favorable conditions, but they also lack the specialized elytral structures that prevent the plants from being rubbed off the back of their host even during beetle copulation.

Within these minute gardens are aquatic protozoans that feed only during periods of high moisture, such as dew, fog, or rain. Nematode worms have also been encountered within these gardens, along with bark lice and mites. One species of mite is strictly a parasite of the weevil itself, extracting precious bodily fluids through the thick exoskeleton of the beetle. An entire family of blind oribatid mites are found exclusively and in abundance on the beetles, with as many as sixty individuals observed just on the surface of the fungal and lichen growth. The plant growth may be initiated by spores that are wind-borne or transported by mites and other animal cohabitants of the epizoic plants. Thick waxy secretions found in the depressions of the pronotum and elytra of the weevils may also have a role in encouraging plant growth. Mating pairs of beetles undoubtedly exchange epizoic flora and fauna as well.

How these weevils benefit from carrying all this flora and fauna about on their backs is unknown, although from a human perspective, camouflage seems likely. From the viewpoint of the gardens and their inhabitants, the beetles provide another niche where space to live and grow is clearly at a premium.

THE HARLEQUIN EXPRESS   The nooks and crannies of the coleopteran exoskeleton provide refuge for a variety of other animals, some of which appear to be specifically adapted to these beetle habitats and, as in the example that follows, depend almost totally on their hosts for transportation to new foraging and breeding sites.

One of the most striking of the neotropical long-horned cerambycid beetles, *Acrocinus longimanus*, is so named because of the extremely long front legs of the males. [ILLUSTRATION V] The common name, harlequin beetle, is a reference to the gaudy pattern on the elytra of both sexes. Despite this conspicuous color pattern, harlequins are perfectly adapted for hiding among the massive, mottled, lichen-covered trunks of hardwood trees, such as *Ficus*. Adult harlequins are widely distributed, feeding on sap and laying their eggs on dead tree trunks of figs and jak fruit. Active both day and night, after dark these beetles are occasionally attracted to light.

The body of the harlequin is home to a community of tiny arachnid passengers that resemble tailless scorpions, which live tucked away beneath the gaily colored elytra. These minute denizens are pseudoscorpions, *Cordylochermes scorpioides*, and although they lack the poisonous sting of scorpions, they are able to subdue their even smaller prey with poison secreted by glands in their claws. All life stages of pseudoscorpions may be found just beneath the bark in tree trunks, living among the sawdust rendered by the activities of the harlequin larvae and other wood-boring insects. Dispersal might be a daunting task for these small, wingless creatures, if not for their coleopterous transporters. Although other pseudoscor-

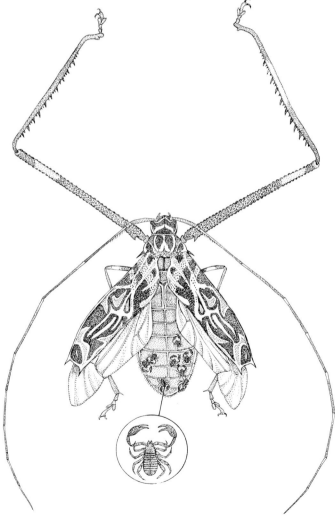

[ILLUSTRATION V]

The Central American harlequin beetle, *Acrocinus longimanus*, provides transportation for pseudoscorpions in search of food and mates. The pseudoscorpions attach themselves with silken safety threads secreted by their pincerlike claws.

pions hitch rides with a variety of beetles, *C. scorpioides* has evolved behaviors that are suited specifically for life among the harlequins.

Why do pseudoscorpions seek out harlequins? Early researchers suspected that pseudoscorpions feed on mites that also inhabit the beetle, but careful observations debunked this hypothesis. Pseudoscorpion individuals that inhabit wood were found to be often in better nutritional condition than their beetle-riding counterparts. Pseudoscorpions use harlequin transport not as sources of food, but as a means to move out of old, decayed logs and onto newly fallen trees. The galleries left behind by wood-boring insects within the tree provide pseudoscorpions with an abundance of food in the form of fly and beetle larvae. Yet the feeding activities of the harlequin larvae quickly reduce prime pseudoscorpion habitat to sawdust. Responding to chemical cues and vibrations, mature pseudoscorpions converge on beetles that are feeding, mating, or laying eggs on their crumbling homes and climb aboard. The tip of the harlequin's abdomen serves as a boarding gate, which the nearly blind pseudoscorpions almost unerringly find. To avoid falling off while the beetle is in flight, pseudoscorpions attach themselves to the beetle by silken safety threads secreted from glands in their claws. When the beetle lands at a suitable log, the pseudoscorpions send out another strand of silk and slide off their transport and onto their new home.

Researchers suspect that, in addition to transporting pseudoscorpions, harlequins serve as mobile mating stations. Big pseudoscorpion males, who compete for space on the beetle's abdomen, tend to stay on board longer, while smaller males and inseminated females rapidly disembark upon arrival at their new home. As the beetle continues its search for mates and sites to lay eggs, new female and male pseudoscorpions climb on board to begin their search for mates and logs with plentiful food supplies.

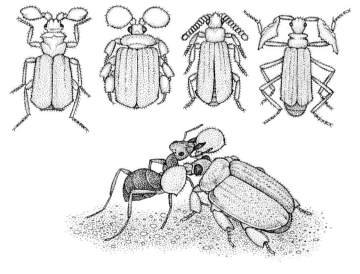

Paussine carabid beetles provide one of the most striking examples of myrmecophily, as evidenced by their strikingly modified antennae. The antennal segments are reduced in number and fused into a single club that can be used by the ants as handles. Each antennal club is packed with gland cells secreting substances which enable the beetles to become an integrated member of the host ant colony.

BEETLES OF THE SOCIAL INSECT'S BESTIARY: MYRMECOPHILES AND TERMITOPHILES

Representatives of many beetle families are found in and among the nests of social insects, exploiting yet another niche and contributing to their overall diversity. Some beetles are simply opportunists, taking advantage of the concentration of resources made available by the activities of ants, bees, wasps, and termites. But many species of beetles are considered to be symbiotic with other social insects, having managed through chemical and/or tactile mimicry to be accepted by their hosts and integrated into the hosts' social system. Beetles that engage in symbiotic relationships with ants, termites, bees, and wasps are known as **myrmecophiles, termitophiles, melittophiles,** and **sphecophiles,** respectively. Both melittophiles and sphecophiles are relatively rare among the Coleoptera.

Of the carabid beetles, the paussines provide a striking example of myrmecophiles. The antennal segments of paussines are reduced in number and fused into a single club. [ILLUSTRATION W] This antennal club is often used by ants as they carry the beetles about the nest and is filled with gland cells that secrete substances to assist with integration into the ant colony. Although paussines are usually collected at lights at night, they have been taken from the nests of a variety of ants, and occasionally, termites, although the extent of their relationship with the latter remains unclear. *Paussus favieri* has been observed in artificial ant nests feeding on larvae, pupae, and parts of adult ants. In addition to their antennal and thoracic glands, paussids bear defensive glands at their hind end that can release caustic chemicals with an audible chirp, staining the fingers of collectors.

Some ptiliid beetles are myrmecophilous. Several genera are specialists living with *Eciton* and other army ants of the Western hemisphere, following the pheromone trails of their hosts. Some myrmecophiles feed by scraping off surface oil and other materials found on the integument of ant larvae, pupae, and occasionally, eggs. For reasons not clearly understood, the ants appear completely indifferent to the feeding activities of these unusual beetles.

Tactile mimicry, that is having the external feel of a member of the colony, may be facilitated by a physical resemblance of the myrmecophile or termitophile to its host (see the discussion of physogastry in Chapter 2). Thus, the hosts themselves probably select ant and termitelike bodies, since ants and termites tend to identify one another by sensing with their antennae. This type of physical mimicry, known as **Wasmannian mimicry**, is most common among members of the family Staphylinidae. The ant- and termitelike forms of these beetles, which amaze human observers, fool their hosts, who go about their daily tasks within the light-deprived confines of their nest. Among antlike staphylinids found above ground marching among columns of army ants and exposed to a full range of predators, however, the physical mimicry of ants is probably an example of Batesian mimicry. In fact, *Ecitomorpha nevermanni* not only looks like an army ant; it is colored like one as well and is able to match the geographic color variation of its host, *Eciton burchelli*.

The larvae of the staphylinid *Atemeles pubicollis* live with the European mound-making wood ant, *Formica polyctena*, integrating themselves into the colony first by imitating the pheromone released by the ant larvae, thereby stimulating larval grooming behavior in the adult ants. Further, by imitating the tactile begging behavior of the ant larvae, the larval *Atemeles* encourages the ants to feed it. The larval beetles spend their summers with *Formica*, but as adults, they move into the nests of *Myrmica* ants, which maintain their brood-keeping and food supplies through the winter, while their spring hosts do not. In spring, *Atemeles* return to the *Formica* nests to mate and lay their eggs. [ILLUSTRATION X]

Most adult and larval histerid beetles prey on arthropods in a variety of habitats. They can be found on trees, decaying plants, dung, carrion, sap flows, bird and mammal nests, and other habitats that are attractive to their prey. However, a large and diverse group of histerids consists of true myrmecophiles and termitophiles. Although most are scavengers in and around the nests, a few have successfully adapted to the specialized conditions of living with ants and termites, approaching the status of true guests.

Many aphodiine, cetoniine, and dynastine scarabs are associated with ants and termites. *Martinezia dutertrei*, an aphodiine scarab often referred to in the literature as *Myrmecaphodius excavaticollis*, can establish a myrmecophilous relationship with more than one species of ant, but not by producing an imitation of the ant's pheromones. *Martinezia* establishes its relationships with ants by absorbing specific hydrocarbons, the scent of its host, into its exoskeleton. Probably introduced from South America with one of its hosts, *Solenopsis invicta* or *Solenopsis richteri*, *Martinezia* has now been found in association with three ants indigenous to North America and

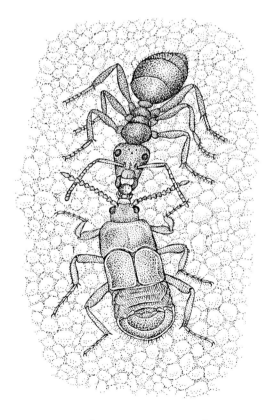

[ILLUSTRATION X]

There are several examples of staphylinid beetles which live with ants. The European *Lomechusa strumosa* is shown here soliciting food from a worker ant, *Formica sanguinea*.

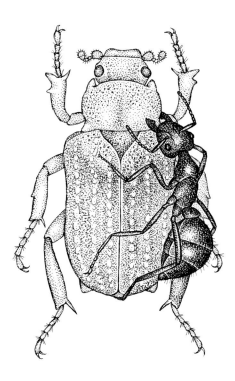

another imported species. Adult beetles move freely about the nest, obtaining food directly from the ants through **trophyllaxis**, an exchange of disgestive fluids between guest and host, preying on ant broods, scavenging dead workers, or stealing food collected by the ants. The beetles continually disperse to other colonies throughout the year. Lacking morphological mimicry and chemical defenses, the scarab is able it integrate itself into its new home by remaining passive, even while under attack, long enough to acquire the colony smell directly through contact with its new hosts.

The North American cetoniine *Cremastocheilus* is a predaceous myrmecophilous scarab. The thickened exoskeleton, protected mouthparts, and thoracic structures packed with setae, which are assumed to exude substances attractive to ants, are all adaptations to a life in close association with ants. [ILLUSTRATION Y] Adult beetles are alternately introduced into and rejected from the nest by their ant hosts. Inside the nests, *Cremastocheilus* may be persecuted or left unmolested to feed on the brood of ants. The larvae are usually found in association with vegetable materials within the nests of ants and rodents. The larvae lack any sort of glands for secreting substances to appease their hosts, but like *Martinezia*, they may be able to absorb nest odors directly into their cuticle.

BEETLE DEPENDENTS Larval lebiine and brachinine carabids and aleocharine staphylinids are specialists that actively seek out and consume the pupae of chrysomelids, gyrinids, and flies. Species of meloids, clerids, and rhipiphorids are known to be parasites of ground-dwelling insects. Meloid larvae attack broods of ground-nesting bees or egg pods of grasshoppers. The clerid *Trichodes* has been found in the nests of bees and reared from the egg pods of locusts. The biology of rhipiphorids is poorly known, but larvae have been found in association with cerambycid larvae, ptiliid beetles, solitary and social wasps, and cockroaches.

Mammalian rodents harbor many ectoparasites, supporting the greatest diversity of biting and sucking lice. Thus, it seems logical that this group of animals would also be the most likely to support a small, yet unique group of beetles that prefer mammals as habitats. The leiodid *Leptinus testaceus* is often taken into the fur of shrews, mites, and moles. A relative of this beetle, the louselike *Platyspyllus castoris*, is found in both larval and adult form on its host, the beaver. These and other beetles have several features that are characteristic of lice and that consequently have occasionally brought into question their placement among the Coleoptera.

One of the most unusual relationships is that of amblyopinine staphylinids with rodents throughout the cool temperate forests and cloud forests of Central and South America. Conjecture and unsubstantiated reports led to the widespread belief that these beetles were blood-feeding parasites. To study the problem, investigators trapped live rodents and kept them in captivity so that both the beetles and their hosts could be observed. Each rodent had from one to thirteen beetles attached to its ears, neck, or elsewhere on the head.

The beetles use their large mandibles to grasp clumps of hair. Several weeks of observation revealed no indication that they feed on the skin, blood, or other secretions of the host. The rodents showed no aversion to having the beetles crawling around their eyes and whiskers and attaching themselves to the rodents' head. These observations indicate that the beetles are probably not parasites at all. In fact, with the exception of being wingless, amblyopinine beetles do not possess the typical characteristics of most parasitic insects, such as flattened bodies and modified mouthparts and tarsi for grasping fur. But like parasites, some species of these beetles show a remarkable degree of host specificity.

This study revealed that the beetles occupy their hosts only at night, spending their days exclusively in the nest. Amblyopinines, like other staphylinids, are predators, feeding on the ectoparasites of their host, such as fleas and mites. The beetles benefit by being transported by their hosts, ensuring that they always have a fresh supply of food as the rodent moves from nest to nest. The larvae of these beetles are still unknown.

{IT'S OUR LIFE TOO}

WE have shown that each species of beetle undergoes a life cycle that specifically adapts it to a rich environment characterized by dynamic conditions and relationships, but we have only begun to touch on the most conspicuous of beetle biologies. The lives of the vast majority of beetles probably will remain in obscurity or be lost altogether in the wake of our ignorance of and rapacious appetite for land.

The greatest impediment to our understanding of beetles and their multiplicity of interrelationships, aside from a lack of financial support, seems to lie in our species-by-species approach to their study. This approach can be likened to attempting to know a symphony by studying it one note at a time and expecting to carry the tune. We must persist in our efforts to observe these marvelous creatures on their own terms, within their natural environment. It is essential that we preserve as much habitat as possible, if for nothing else, to ensure that beetles continue to exist so that they can be studied by future generations of scientists, who will undoubtedly be better equipped to unravel the details of their lives. The rewards for such efforts will surely be great, revealing strategies employed by beetles that have been evolving for millions of years and may provide essential elements to a blueprint for our own survival. [PLATE 72]

[PLATE 72]

Brilliant colors and striking patterns, such as those seen here on the Central American cetoniine scarab, *Gymnetis stellata,* are thought to help break up their image, so that predators looking for beetles will not recognize them as such. *Specimen courtesy of the Natural History Museum of Los Angeles County.*

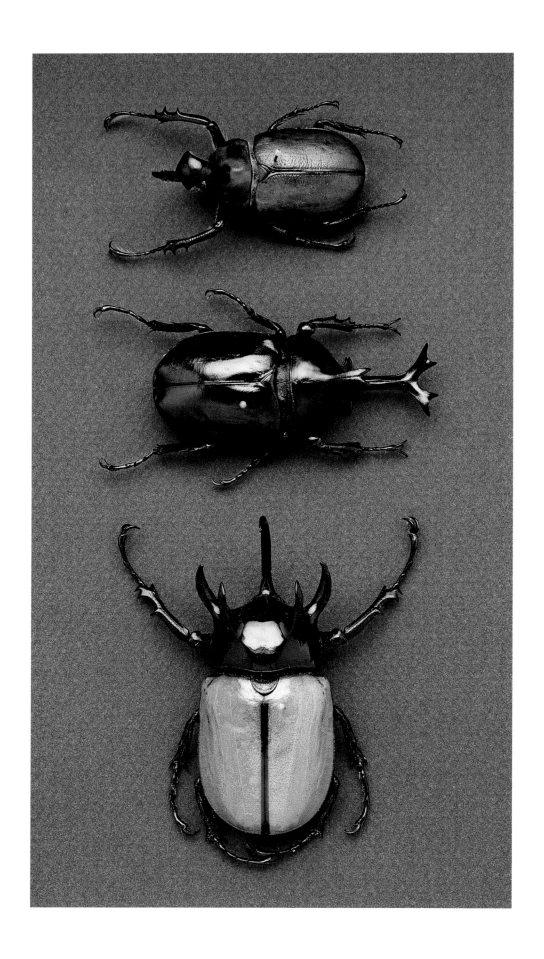

five

BEETLES AND HUMANS

*Whenever I hear of the capture of rare beetles, I feel like an
old war-horse at the sound of a trumpet.*

·

Charles Darwin

I N a sense, beetles have dramatically affected the way we view the world. Even the etymology of the word *beetle* in many languages reveals a relationship between humans and insects that was forged through a combination of antagonism and fear and the ultimate realization that beetles often compete with us for food and space. The incredible diversity and overwhelming numbers of beetles are constant reminders of our subordinate role in nature.

The sacred scarabs of Egypt were worshiped by humans at least as early as 2000 B.C. Beetles were and are an important food source to peoples throughout the world. Their flickering lights, their bewildering palette of color, their diverse structure, [PLATE 73] and the pharmaceutical properties of their body fluids have long captured our attention and imagination. For some of us, beetles are the basis for a lifelong quest. Few people realize that Charles Darwin himself was an irrepressible beetle collector and that his passion for beetles would eventually deflect him from his clerical studies, ultimately leading him to develop his theory of natural selection. At first, though, human interest in these creatures was of a more practical nature, as evidenced by the names that were given to them.

{WHAT'S IN A NAME?}

T HE naming of beetles is hardly an exact science. Many of the examples we will discuss are not universally accepted by scholars. However, the nomenclature used for beetles throughout the world reflects primarily their form and presumed function within the ecosystem. The fact that ancient writers were not careful to distinguish different species of beetles has resulted in some confusion about the derivation of early popular beetle

[PLATE 73]

The diverse colors and structures of beetles have long fascinated scientists and amateurs alike. Charles Darwin was an irrepresible beetle collector in the years leading up to and during the the voyage of the the H.M.S. *Beagle* and it was this passion, in part, that deflected him from his clerical studies.

Horned scarabs such as these often spark an interest in insects. Top to bottom: *Golofa pizarro*, Mexico; *Allomyrhina dichotomus*, China; and *Eupatorius gracilicornis*, Malaysia.
Specimens courtesy of Paul McGray.

names and about whether the appropriate quality has been attributed to the correct group of beetles.

The presence of chewing mouthparts in the Coleoptera strongly influenced derivation of the word for these insects in several languages. The English *beetle* is traced from the Middle English *bityl* or *betyl* and the Old English *bitula*, meaning "little biter." The English name *chafer*, now applied almost entirely to plant-eating scarabs, is derived from the Old German *Kevar*, which is the source of the modern word for beetle in that language, *Käfer*.

Sometimes the behavior of a beetle may have influenced its name. Another word for beetle that is Germanic in origin, *Webila*, denotes a movement back and forth, flickering, swarming, or teeming. The application of this name no doubt became appropriate when humans observed the activities of beetles that infested their food. At one time *Wibel* was used to refer to all beetles, but this term was eventually replaced by the word *Käfer*, although usage of the English form, *weevil*, still persists today. [PLATE 74]

The ancient Greek terminology for beetles originates from the understanding (or lack thereof) of the time of the beetle's biology. The beetle genera *Cantharis*, *Melolontha*, *Buprestis*, *Cerambyx*, and many others, are derived from ancient Greek and/or Latin terminology. *Kantharos* (Latin *cantharus*) is a wide-bellied vessel with handles. Some scholars have suggested that, at least in pejorative or humorous usage, the term *kantharos*, used to describe some dung beetles, is derived from the word *kanthon* (ass), in the erroneous belief that beetles were produced from the rectum of animals!

The term *Melolonthe* has proven to be quite problematic for etymologists. *Melolonthe* has been described by some authors as a combination of the words *melon* ("sheep, a grazing animal") and *olonthos* ("wild fig"), no doubt referring to the fact that some melolonthine scarabs in the Mediterranean region prefer to feed on leaves of the fig tree.

The metallic wood-boring beetles, or buprestids, derive their name from the Greek *bous* and *prestis* (a form of the verb *prethein*, "to blow up"), which together mean "cow inflater." It was thought that these insects, while lurking in the grass, were consumed by grazing cattle, causing the cows to swell and burst. In reality, the ingestion of rove beetles (staphylinids) or blister beetles (meloids) is probably what caused the gastrointestinal distress of bovines.

In the Middle Ages, certain species of coccinellid beetles were dedicated to the Virgin Mary and named *beetles of Our Lady*. As time progressed, *ladybird beetles*, *ladybirds*, or *ladybugs*, became popular names with English-speaking children. In fact, *ladybird* appears as a term of endearment in Shakespeare's Romeo and Juliet The popularity of the ladybird beetle continues today, as they are featured in nursery rhymes and verses. In Scandinavia, ladybird beetles are called *nyckelpiga*, "Our Lady's key-maid," or *jung fru* or *Marien-kafer*, "lady beetles of the Virgin Mary," and in France is *bêtes de la Vierge*, "animals of the Virgin." In England, they are variously known as ladybirds, ladybugs, lady flies, and May cats.

In English-speaking countries, some scarab beetles, particularly those that crash about lights at night in the spring, are called May beetles. The Swedish name is *maibagge*; in Germany they are referred to variously as *Maikäfer*, *Maikawel*, *Maikiefer*, or *Maikobbelt*. In Italy, the clumsy flight of these beetles at twilight has earned them the name *buffone*, meaning "buffoon" or "clown."

ONCE named, beetles became an inextricable part of our lives. Armed with a vocabulary wrought from common experience, but lacking the knowledge of insect metamorphosis and sexual reproduction that we often take for granted today, our ancestors began to discuss the emergence and behavior of some beetles that must have appeared to them as truly magical. These observations, bolstered by perceptions based on fact and fiction, undoubtedly are what led some ancient civilizations to consider the activities of beetles to be metaphors of the human world. Interestingly, however, while some beetles were thought to symbolize the activities of deities, others were despised as evil.

Although featured in ancient religions and mythologies throughout the world, nowhere did beetles, particularly the dung scarab, play as significant a role as in the civilization of ancient Egypt. [ILLUSTRATION Z] Egyptians believed that the activities of the dung scarab represented their own world in miniature. Every year these industrious creatures buried balls of dung, and every year more beetles suddenly emerged from the ground. In ancient Egypt the sun god, Ra, was symbolized as a great scarab, rolling the sun, like a dung ball, across the heavens of a universe with Earth at its center. The burying of the dung ball came to symbolize the rising and setting of the sun.

The scarab beetles, like Ra, were thought to be male and born not from the union of two sexes, but from inanimate matter. These male beetles were believed to deposit their seed in the dung ball to produce future generations of beetles. It has been suggested that the ancient Egyptians knew of the scarab's metamorphic process and that the pupae, formed within the buried dung ball, inspired human mummification within underground chambers. Yves Cambefort of the Musée National d'Histoire Naturelle in Paris, highlighting the parallels between the reproductive biology of the dung scarab and mummification, suggests that the dung pats of cattle and chambers created within and below them by dung scarabs may have inspired early Egyptians to construct the pyramids, which also have many chambers, to prepare their dead for the afterlife.

The symbolism of the scarab progressed in the New Empire, where it was interpreted as the embodiment of the god of creation, Cheper, one of the outward manifestations of Ra. Later the scarab came to represent the souls that were to unite with Cheper.

Images of sacred scarabs appeared everywhere in ancient Egypt, from art inside burial chambers to hieroglyphs. Carved scarabs bore religious inscriptions from the *Book of the Dead* or were wished upon for good luck, health, and life and placed in the tombs of the earliest dynasties to ensure the immortality of the occupant's soul. Heart scarabs were placed on or near the chest of the mummy and bore inscriptions admonishing the heart not to bear witness against its own master on judgment day. Scarab mythology was so pervasive in ancient Egypt that carved scarabs were worn as good luck amulets by the soldiers of the occupying Roman forces for their presumed protective powers in battle. Today, many people wear scarabs as a symbol of good fortune, not unlike the shamrock or rabbit's foot, but few appreciate the origin of the scarab's mythological significance.

Beetles also appear in ancient myths and legends of Greece, symbolizing everything from industriousness to insignificance. A Greek fable tells of

[ILLUSTRATION Z]

The ancient Egyptians believed that the activities of the dung scarab represented their own world in miniature. The ancient Egytptian sun god, Ra, was symbolized as a scarab, rolling the sun, like a dung ball across the heavens of a geocentric universe.

an eagle that invades the asylum given by a dung beetle to a hare. The eagle eats the hare, and in revenge, the beetle takes to destroying the eggs in the eagle's nest. The eagle, as the bird of Zeus, obtains permission to lay its eggs in the lap of Zeus. The beetle flies over Zeus and drops dung on the god's lap, causing Zeus to brush the dung away and, in the process, to destroy the eagle's eggs. In another fable a beetle emerges from its dung heap and is overcome with envy by the sight of a soaring eagle and tries to emulate the bird. The beetle soon becomes exhausted, landing far from home, and expresses its longing for the old dung pile. A dung scarab opens Aristophanes' play *Peace*. Trygaeus, concerned about whether the Peloponnesian War can be brought to a halt by normal methods, flies to Zeus for advice, not on the back of Pegasus, but on the back of a large dung beetle. The use of dung scarabs in ancient Greek fables and plays underscores the comic extravagance of the time and allows for the insertion of plenty of scatological humor!

In medieval Christianity, the scarab was thought to symbolize the sinner. The dung scarab's natural and necessary association with excrement helped encourage its association by humans with foulness and wickedness. This association no doubt led to the later belief that the dorbeetle, *Geotrupes stercorarius*, was a symbol of bad luck. In parts of Austria, Germany, and Sweden, this beetle was known as the witch beetle and was thought to be associated with the devil. Killing the beetle would bring bad luck, but if a farmer rescued a dorbeetle lying helplessly on its back, he could save his crop and home from destruction by the elements.

The European stag beetle, *Lucanus cervus*, had a bad reputation for fire raising. [PLATE 75] The large mandibles of the male were thought to carry glowing embers to a thatched roof. Stag beetles were also believed to attract lightning. It has been suggested that this association resulted from the habit of stag beetle larvae to inhabit isolated oak trees, which may frequently have been struck by lighting. Today, the head and mandibles of male stag beetles are sold in Turkey as good luck charms to ward off the evil eye.

In Irish mythology, the devil's coach horse, *Staphylinus olens*, was supposed to have the power of killing by a look, or cursing hapless human victims by simply raising the tip of its abdomen. The connection with the devil is derived from the belief that the beetle consumed the bodies of sinners. It was also thought that the beetle would appear in the hand of someone who had dealings with the devil.

Although most of the readily available information about the religious and mythological significance of beetles deals with European culture, beetles figure prominently in cultures elsewhere in the world. In the mythology of a South American tribe, a beetle is the creator of the world, who, from the grains of earth that were left over, also created men and women. The bizarre giraffe weevil, *Lasiorrhyncus barbicornis*, is one of the most grotesquely shaped weevils in New Zealand. Because of its striking resemblance to the shape of their canoes, the native Maori dubbed this weevil *Tuwhaipapa*, the god of the new-made canoe.

{THE ARTISTIC BEETLE}

LONG before beetle references appeared in literature, they could be found in art. Because of their diminutive stature and supposed insignificance, however, they do not appear as commonly as other creatures.

[PLATE 75]

In medieval Europe the stag beetle, *Lucanus cervus*, was claimed to have been responsible for starting fires by carrying glowing embers with their exaggerated mandibles and depositing them on thatched roofs. Today, the head and mandibles of these and other stag beetles are sold in Turkey as good luck charms to ward off the evil eye. *Specimen courtesy of Maxilla and Mandible.*

[PLATE 76]

The dazzling array of colors and forms of beetles have inspired the use of their bodies as jewelry. Some of these species of cassidine chrysomelids are popularly used in necklaces, earrings, pins, and other human adornments. *Specimens courtesy of the Natural History Museum of Los Angeles County.*

[PLATES 77 AND 78]

The durability and color of the beetle exoskeleton has inspired
their use in jewelry. The metallic elytra of this species of
buprestid beetle, *Euchroma gigantea*, were used in this
Amazonian Indian necklace made near Yanaomo in Loreto
Province of Brazil. *Courtesy of the Natural History Museum of
Los Angeles County.*

Beetles have been depicted in vase paintings, porcelain statuary, precious stones, glass paintings, sculptures, jewelry, coins, and illustrated manuscripts.

Beetles have often formed the central theme for other works of art. One of the most notable examples of beetles in art is Albrecht Dürer's watercolor of the stag beetle (1505), which is the oldest known work whose subject can be readily identified to a particular species, *Lucanus cervus*. Fireflies have always held a special fascination for the Chinese and Japanese and have often appeared in their art. A Japanese woodcut by Hosoda Eishi (1756–1829) depicts three women on a river, surrounded by a swarm of fireflies. One of the women is trying to catch a firefly with a fan, while another has a cage filled with already caught luminescent beetles. The capture of fireflies may take on an almost erotic quality, as rendered by Kiyomitsu (1735–1785). In this painting, a young woman, having just completed her bath and dressed in an open, transparent gown, is attempting to capture fireflies with a fan. The fireflies are shown to be blushing! The French artist Eugene Seguy's art deco insect portfolio, created in the 1920s, includes several striking examples of Coleoptera.

Sacred scarab ornaments may have appeared with other insect jewelry in Egypt as early as 1600 to 1100 B.C. As symbol of the sun god, Cheper, carved scarabs were not only placed with mummies, but adorned royalty as well. Carved beetles were common on amulets, necklaces, headdresses, and pendants. Egyptian warriors wore scarab rings as a sign of masculine valor. Images of beetles were carved from stone or fashioned from glass or metal. Occasionally other durable materials were used, such as ivory, wood, and nutshells. The Japanese carved insect pendants, or *netsuke*, in a variety of insect forms, including one that resembled the larva of a weevil. The manufacture of insect jewelry continues today; some catalogues price scarab and ladybird beetle jewelry at as much as several thousand dollars. In the Middle Ages, casting from nature had already been known for centuries. Casts of stag beetles were used to adorn decorative boxes, bowls, and writing sets.

The dazzling array of colors and forms of beetles have not only provided artistic inspiration; the bodies of the beetles [PLATE 76] themselves have been incorporated into jewelry for hundreds of years. The durability of their chitinous bodies has encouraged their use by ancient and modern designers. The elytra are used most often, but the horn, mandibles, and legs also have been prominently featured. The metallic elytra of buprestid beetles are used around the world. The native Jivaro peoples of Ecuador used the elytra of buprestid beetles to make earrings. Other native South American tribes use the elytra of the giant *Euchroma gigantea* for necklaces, head ornaments, and other decorative pieces. [PLATES 77 AND 78]

Click beetles of the genus *Pyrophorus* (from the Greek for "fire bearer"), also known as headlight or lantern beetles, are the only click beetles that produce light. The intense glow comes from two luminescent organs on the upper side of the prothorax. Encountering this beetle for the first time in the dark, one might think that it was two lampyrid beetles flying in tandem. *Pyrophorus* species inhabit the West Indies and range from the extreme southwestern United States southward to Brazil. Reportedly, nocturnal travelers could tie a pair of these beetles to their big toes to light their path. By placing several such beetles in a perforated gourd, the nineteenth-century German naturalist and explorer Alexander von Humboldt was able to fashion a reading lamp. In Mexico, Indians were reported to smear a paste on their bodies made from the glowing beetles as a joke to scare those not familiar with the prank. At a certain dance women adorn themselves with adult beetles. The beetles are fitted on a hairpin or placed in gauze bags and arranged on the blouse or shirt. The larvae are also luminescent, as well as predaceous, feeding on scarab and other beetle larvae. In the Caribbean, lampyrids were used in a similar fashion, placed in gauze sacs and worn as hair ornaments.

The best-known example of living insect jewelry in the Western hemisphere comes from Mexico and Central America, with the use of the zopherid *Megazopherus chilensis*. The beetle ranges from southern Mexico to Venezuela, where it may be found rambling about the bark of dead trees, presumably feeding on fungal hyphae (threads). In Yucatán, this beetle, popularly known as the *ma'kech*, is decorated with brightly colored glass beads, fixed to a small chain or tether, and pinned to clothing as a reminder of an ancient Yucatecan legend. [PLATE 79] A young Mayan prince was turned into this beetle by the moon goddess in order to prevent his capture by the guards of his lover. The maiden was so impressed by the prince's cleverness in overcoming the many obstacles in the path to their love that she uttered the phrase "ma'kech," which meant "thou art a man," referring to the prince's courage. *Ma'kech* also means "does not eat," a reference to the prince's ability to endure prolonged fasts during his ordeal—a character also attributed to the zopherid. The zopherid's thick exoskeleton affords it some protection from desiccation, and if properly cared for, the tethered living jewels may survive several months on a diet of rotten wood, cereal, and apples.

In 1991 the insect zoo at the Natural History Museum in Los Angeles received a call from the Pasadena chapter of the Society for the Prevention of Cruelty to Animals. They had received one of these bejeweled coleopterans, resplendently encrusted with rhinestones. Apparently someone had noticed the beetle crawling through the gutter of a residential area and, because of its bright colors, thought that the beetle was poisonous and notified the police. The officer noted the chain glued to the body of the beetle and suspected that it might be a case of animal abuse and alerted the SPCA. Ideal grist for the media mill, this story was covered nationally and internationally. Upon learning that the Natural History Museum was starting an insect zoo, the SPCA turned the beetle over to the museum, where it lived for nearly a year.

{FRIENDS AND FOES}

DESPITE the artistic inspiration provided by the color and form of beetles, our relationship with them has always been driven primarily by practical concerns. A small number of beetle species—feeding and laying eggs on crops, pastures, timber, and stored products—may be of considerable economic importance. One-third to one-half of all food grown for human consumption is lost to insect damage. Plant-feeding beetles, particularly those that concentrate their activities on legumes, tomatoes, potatoes, melons, gourds, and grains, are one of humanity's greatest competitors for food.

The subterranean larvae of melolonthine and dynastine scarabs separate the upper photosynthetic structure of grasses from the roots, causing

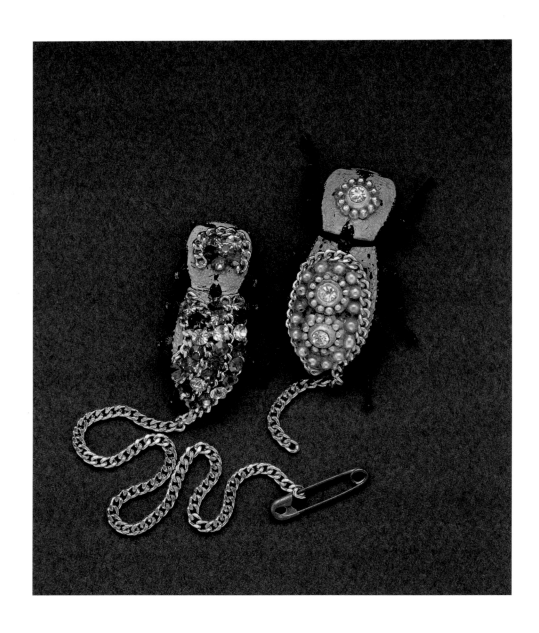

[PLATE 79]

A zopherid, *Megazopherus chilensis*, which is distributed
throughout Central and South America, is used in Yucatan as
living jewelry. Popularly known as the *ma'kech*, the zopherid
is decorated with brightly colored glass beads, fixed to a small
chain or tether, and pinned to clothing as a reminder of an
ancient Yucatecan legend. *Courtesy of the Natural History
Museum of Los Angeles County.*

[PLATE 80]

Wood boring bark beetles and cerambycids, such as the eucalyptus borer, *Phoracantha semipunctata*, usually attack trees that are already stressed from lack of water and proper nutrition. This Australian species now threatens ornamental eucalyptus in California. *Specimen courtesy of the Natural History Museum of Los Angeles County.*

dead spots in lawns that most home owners tend to attribute to lack of proper maintenance or the territorial behavior of a neighbor's dog. The wood-boring bark beetles and cerambycids usually attack trees that are already stressed from lack of water and nutrition. [PLATE 80] With their defense systems impaired, these trees may succumb to diseases that are transported on the bodies of the beetles. Stored-product pests are often associated with grains, cereals, pastas, dried fruit, leather, wood, dried meats, furs, and tobacco.

The sputtering and popping coming from a cigar may be the result of the rapid expansion of body fluids and subsequent violent rupture of surrounding exoskeleton of the larval or adult tobacco beetle, *Lasioderma serricone*. The drugstore beetle, *Stegobium paniceum*, may be found dining on a concentrated diet of cayenne pepper. Confused flour beetles, *Tribolium confusum*, have been found infesting bales of marijuana. (The species name for this animal, however, is derived not from its mental state after ingesting the leafy contraband, but from its misleading similarity to another beetle.)

PLAYING IN THE COTTON BOLL
One of the greatest battles humans have ever waged against a beetle was directed at the boll weevil, *Anthonomus grandis*. Large-scale planting of any single crop, referred to as **monoculture**, is an inherently unbalanced ecosystem. The slightest perturbations caused by adverse weather, physiological or behavioral changes of the crop or pest species, introduction of an exotic pest species, or the synergistic effects of any or all of the aforementioned maladies can quickly result in the financial ruin of a farmer. The extensive monoculture of cotton was the major financial investment in the United States for much of the nineteenth century, and cotton fields blanketed much of the Old South. The total economic dependence on cotton in this region was the primary ingredient in a recipe for certain financial disaster.

The origins of cotton in the Western hemisphere are not clear. Wild cotton, of the genus *Gossypium*, is found in arid regions of the tropics and subtropics of Africa, Asia, Australia, and the Americas. In 1786 the seed of *Gossypium barbadense* was transported from the West Indies to South Carolina, where it developed into an annual crop. The annual variety became known as Sea Island cotton and, at the time, was superior to the perennial variety, which allowed the accumulation of insect pests to epidemic proportions. Even the less susceptible Sea Island cotton, however, was almost completely wiped out in the United States when the boll weevil moved across the cotton belt from Mexico near the end of the nineteenth century.

Boll weevils were first discovered by L. A. A. Chevrolat in Veracruz, Mexico, in 1830. Chevrolat sent the specimens to the Swedish zoologist C. H. Boheman for description. The boll weevil was not detected in the United States until 1892, where it was reported to have crossed the Rio Grande near Brownsville, Texas. In 1920 it was reported in Arizona. By 1922 the boll weevil had infested all the cotton-growing areas in the southeastern United States and most of Texas, spreading at the rate of about 60 miles a year. The boll weevil was encountered in Venezuela in 1949, Colombia in 1951, California in 1982, and Brazil in 1983. Today, in the United States, the boll weevil occupies nearly all the cotton-growing belts in the United States, except the Central Valley of California.

Adult boll weevils attack the cotton plant by chewing cavities into the

The widespread destruction wrought by the boll weevil had a positive impact on the citizens of Enterprise, Alabama, forcing them to diversify by planting other crops.

young blossom, which is called a square. [ILLUSTRATION AA] They then lay their eggs in these cavities early in the growing season. Later in the season the cotton boll is the preferred egg-laying site. On hatching, the larvae feed within the cavity begun by the adult beetle. After a short pupal period, the freshly emerged beetles chew their way out of the cavity. Each generation takes about twenty-five days, and as many as ten generations are produced each year. The activities of the boll weevil are devastating because either it destroys the bud before the cotton is produced, or it damages the cotton boll to such an extent that the seeds have very little fiber. Damage caused by these beetles in the United States alone is estimated in the tens of millions of dollars annually.

Several attempts to control the boll weevil have met with failure. At the turn of the twentieth century, the Texas legislature tried to establish cotton-free zones to prevent further spread of the weevil. Trap cropping—that is, planting crops prior to the normal planting time to lure overwintered beetles to them for pesticide treatments—have had only mixed success. In the early 1960s a parasitic nematode worm was discovered to infect boll weevils, but it proved to be ineffective in controlling their populations. The kelep ant from Guatemala, which was found to prey on boll weevil larvae, failed to become established in the United States because it could not adapt to the temperate climate. An African parasitic wasp introduced in the 1930s and again in the 1960s also failed to be come established. Chemical sterilants, attractants, repellents, insecticidal baits, and attempts to develop sexually active but sterile weevils through hybridization have all failed as methods of control. Recently, intensified eradication programs have achieved some degree of success using various methods of cultural control, such as varying tilling, planting and harvest times, improved baited traps, the release of sterile males, and quarantines in parts of California, Arizona, and New Mexico.

The widespread destruction wrought by the boll weevil had a positive impact on the citizens of Enterprise, Alabama. Several times the boll weevil devastated Alabama's cotton crop, resulting in a 60 percent loss in yield. Farmers were forced to diversify by planting other crops, such as peanuts, corn, and potatoes. Much to their surprise, the farmers reaped substantially higher profits from these crops than they had realized from cotton. The ravages of the boll weevil thus led to a more diversified and thriving economy, and the citizens of Enterprise erected a monument in honor of the weevil bearing the inscription: "In profound appreciation of the Boll Weevil and what it has done to the Herald of Prosperity. This Monument was erected by the Citizens of Coffee County, Alabama." Cotton has not disappeared completely from the region. Today, thanks to government subsidies, both cotton and boll weevils continue to thrive.

LEARNING FROM OUR MISTAKES

The agricultural community did not immediately embrace the lessons learned by cotton farmers. Although farmers in some regions began to practice a small degree of crop diversification, the vast majority of farmers continued to practice monoculture. Outbreaks of exotic pests were an ongoing problem. Native species of beetles achieved greater pest status as the monoculture destroyed the balance of natural population suppressors, such as limited food supply, predation, and disease. To deal with these pests, farmers came to rely on the insecticidal properties of various chemical concoctions.

The seductive convenience of chemical controls failed to take into account the incredible adaptive abilities of insects. At first, these applications of pesticides seemed to be a panacea for the farmer; unwanted pests were killed by the thousands. But a few individuals always persisted to reproduce, passing along to the next generation their genetically based immunity to the toxins. This unfortunate and, at the time, poorly understood scenario was repeated throughout the agricultural world, resulting in ever-increasing amounts of pesticides being applied to kill fewer pests.

In our hurry to kill these pests, we failed to recognize that many pests become resistant to pesticides and that the persistent nature of many pesticides effects untargeted plants and animals, including humans. Today, total reliance on chemical controls has given way to the practices of the sterilization of pests by radiation, genetic engineering, and the augmentation of existing predators, parasites, and pathogens. It remains to be seen whether these methods will lead to an ecological crisis that parallels our early experiences with chemical controls.

Although many people had warned of impending doom from the overuse and abuse of pesticides, not until Rachel Carson's book *Silent Spring* appeared in 1962 did the general public become aware of the need to restrict and alter our use of pesticides. **Integrated pest management,** the culmination of this awareness, uses a variety of strategies for the control of insect pests, including cultural, chemical, and biological controls.

BIOLOGICAL PEST CONTROL   The role of predator and parasite associations, particularly in relation to plant-eating insects, laid the foundation for biological control. However, the use of insects as control agents is not a new concept. Ancient Chinese texts recount using ants to control stinkbugs, stored-product pests, and citrus pests. Large predaceous carabids and coccinellids were recognized in Europe, and the purposeful introduction of these predators was undertaken locally to curb pest outbreaks.

The modern use of a pest's natural enemies began with the importation of ladybird beetles from Australia into California to battle a particularly pesky relative of aphids and mealy bugs known as the cottony cushion scale, *Icerya purchasi*, on citrus. The cottony cushion scale was first noticed on acacias imported from Australia in Menlo Park, California, in 1868. By 1885 this insect had become a serious pest of citrus orchards throughout the state, threatening the entire fledgling industry. Charles Valentine Riley, entomologist with the Department of Agriculture, suspected that the scale was Australian in origin. In 1887 he sent Albert Koebele to Australia to collect natural enemies of the scale. Koebele sent back three shipments containing several species of coccinellids, including *Rodalia cardinalis*, to California.

These imported beetles became the first exotic insects to be introduced into North America for the purpose of biological control. Daniel Coquillett placed the original 129 beetles on a small infested citrus tree enclosed by a tent on a Los Angeles citrus farm. Within a few months the scales were gone. Thereafter the beetles were released among the surrounding trees, and in just two months more than 10,000 beetles were collected and distributed among citrus farmers throughout southern California.

[PLATE 81]

Dung scarabs, such as this Central American *Oxysternon conspicillatum*, provide an essential service by burying dung. African dung beetles, including both dung scarabs and predatory histerids, have been used as biological control agents in the Australian battle against pestiferous horn flies. *Specimen courtesy of Maxilla and Mandible.*

Within a year the citrus crop had increased nearly threefold, marking the complete recovery of the citrus industry.

The immediate success of the introduction of *Rodalia* led to increased interest in biological control, precipitating the introduction of more than forty-five other species of ladybirds between 1891 and 1982, but only a few of the introductions were successful. To date, 179 species of coccinellids have been introduced, either on purpose or accidentally, into the United States. Twenty-eight of these have become established. The full potential of ladybirds and other beetles as biological control agents may have yet to be realized.

Predatory beetles are not the only species with potential as biological control agents. Many plant-feeding species of beetles may prefer to eat plants that we consider to be weeds and thus may have great potential for controlling such weeds. The leaf-feeding activities of chrysomelids, twig-boring activities of buprestids, and seed-eating activities of weevils have been used to control the spread of noxious weeds throughout the world. The American longhorn cactus borer *Monoleima*, for example, was imported as part of a biological control program in Australia to control the spread of the introduced prickly pear cactus.

Other examples of beetles that have been successfully imported to control noxious weeds are two species of jewel beetles, both introduced into the United States from Europe. *Agrilus hyperici* was imported and released to control the Saint-John's-wort *Hypericum perforatum*, a weedy plant that has become established in the Pacific Northwest. *Agrilus* females lay their eggs on the plant, and the damage caused by the tunneling larvae prevents the reproductive parts of the plant from reaching maturity; thus seed production is reduced and fewer new plants germinate. A second jewel beetle, *Sphenoptera jugoslavica*, has been released in Washington, Oregon, and southwestern Canada to control a rangeland pest known as diffuse knapweed, or *Centaurea diffusa*. *Sphenoptera* larvae adversely affect the reproductive vigor of the weed by tunneling through its root crown.

One of the most intriguing chapters in the history of biological control revolves around the Australian bovine dung problem. Dung beetles, including the excrement-eating, dung-rolling scarabs and the predatory dung-inhabiting histerids, have been used as biological control agents in the Australian battle against horn flies. [PLATE 81] On their arrival "down under" in 1778, English colonists introduced a host of domestic animals, many of which produce large, moist dung pads that were entirely unsuited to the native dung beetles. Dung beetles indigenous to Australia had evolved to use the dry, fibrous dung pellets produced by the native marsupials. As long as cattle were scarce, their dung did not pose a problem. But as cattle populations increased and became more dense, pastures became littered with pads that took up to several years to decay. In fact, termites and weathering, not dung beetles, were responsible for much of the degradation of the dried pads. It was estimated that in just thirty minutes, 6 million bovine dung pads were deposited on the surface of Australia, reducing the amount of consumable pasturage and serving as egg-laying sites for pestiferous horn flies.

In 1963, authorities in Australia decided to begin establishing a wide range of imported dung beetles that would dispose of bovine dung pads.

These introduced beetles needed to be able to thrive in the climatic conditions prevalent in the cattle-breeding areas of Australia. The savanna of southern Africa proved to be climatically comparable to the cattle-raising areas of Australia, and entomologists were soon dispatched to study indigenous African dung beetle communities. They paid careful attention to how each species of dung beetle used dung so that they would import the proper complement of beetle taxa into Australia to ensure that all parts of the dung pad would be decomposed.

Since dung beetles specialize in dung, it was unlikely that they would infest other materials. But there was considerable concern that their release into Australia might also result in the introduction of such devastating cattle diseases as foot-and-mouth disease and rinderpest. The bodies of the beetles were found to be teaming with mites, nematode worms, fungi, and bacteria. The beetles could be treated with specific pesticides and fungicides, but even these treatments would not guarantee the elimination of all harmful organisms.

To reduce the risk of introducing unwanted microorganisms into Australia, beetle eggs in South Africa were removed from their dung balls, washed, placed in sealed containers, and shipped under quarantine to Australia, where they were transferred to "artificial" dung balls made of Australian bovine dung. The first generation of dung beetles was allowed to mate and lay eggs. These eggs were then sterilized on the surface and removed from quarantine to begin the mass rearing program. In 1967 four species of dung beetles were released, one of which, *Onthophagus gazella*, quickly became established throughout the continent. The activities of imported dung scarabs helped reduce the amount of dung available for horn flies to lay their eggs on, and the introduction of the predatory histerids from Africa helped reduce the number of horn fly maggots that were able to complete their life cycle.

In North America, dung beetles have been imported and released in several states to augment native populations of excrement-eating beetles. Unfortunately, little effort was made to determine the efficacy of dung removal by native dung beetles or the subsequent impact that the introduction of exotic species would have on native beetle populations. Preliminary observations in Arizona indicate that the introduced *Onthophagus gazella* has significantly displaced the native species of *Canthon*, *Onthophagus*, and *Copris*. The short- and long-term effects of this displacement remain to be studied.

The future of biological control is promising. Our reliance on chemical controls has given way to the use of biological controls: predators, parasites, and pathogens. In our zeal to introduce these exotic predators and pathogens, however, we must be aware of the possible repercussions on untargeted species. Many biological control agents are specific only to whole genera, families, or orders of insects. Their introduction into the environment could have unknown and devastating consequences for related taxa, particularly those that are considered to be endangered species. Beetle species with a tenuous foothold may end up paying the ultimate price of extinction simply because they share a physiological susceptibility with a pestiferous species that is being combated by the introduction of an exotic, lethal, and nonspecific biological control agent.

Western man, despite his frequent temptation to claim his foodways are based on rational considerations, is no more rational in this than other men, for it makes no better sense to reject nutritious dogflesh, horseflesh, grasshoppers and termites than to reject beef or chicken flesh. —Frederick Simoons

RELIGIOUS admonitions aside, it is quite natural to consider some invertebrates to be delicacies. Mollusks, such as escargot, calamari, clams, mussels, abalone, and scallops, to name a few, are readily consumed throughout the world. Even some arthropods, particularly crustaceans, such as lobster, crab, and shrimp, are a regular part of the human diet, even though they are nothing more than chitinized garbage disposals that consume animal waste and putrid flesh. Yet the mere thought of eating insects, who are among the tidiest of animals, makes most Westerners shudder. Why? Some suggest that we distinguish between invertebrates that share our habitat and those that don't. We seem to relish marine creatures but have clear disdain for terrestrial invertebrates.

Research has established that insects are an excellent source of protein, and they are an important component of the human diet in parts of Africa, Australia, and Asia. Grasshoppers and termites are the most commonly consumed insects, but many beetles are relished as well. Beetles either supplement or are the primary source of protein and fat for the human diet in many parts of the world. Worldwide, the plump, fat larvae of cerambycids and scarabs are probably the most commonly consumed beetles.

In the not so distant past, native peoples of North America consumed insects, particularly the larvae of long-horned beetles. These chitinous sausages, ranging from 7 to 60 millimeters in length, are an excellent source of fat and were removed from dead or dying trees and eaten raw. In California, the robust larvae of the pine sawyer, *Ergates spiculatus*, and the California prionus, *Prionus californicus*, were considered delicacies. Other cerambycid larvae known to be sought and consumed by California indigenous peoples include *Rhagium*, *Xylotrechus*, and *Monochamus*. Infestations of weevil grubs in acorn and pine nut meal not only enhanced their nutty and oily flavor, but also provided an additional source of protein. The adults of one type of striped June beetle, *Polyphylla crinita*, was frequently consumed.

The ancient Romans consumed the larvae of cerambycids and lucanids, after specially fattening them with a spicy meal of flour and wine. The father of the modern study of insect behavior, Jean-Henri Fabre, once served grubs prepared in this manner to guests, who later reported that the flavor of the meal suggested almonds with the slight aroma of vanilla.

Pyronota festiva, a beautiful shiny green melolonthine scarab, is known in New Zealand as the manuka beetle because it is found in summer swarms on the flowers of the manuka plant, *Leptospermum scoparium*. Occasionally, large numbers of these beetles that had become mired along the muddy banks of streams were collected and eaten by the Maori, who called them *kerwai* or *reorepowai*, from *repo*, "mud," and *wai*, "water." The Maori also consumed the grub of the largest beetle found on the islands of New Zealand, a cerambycid named *Prionoplus reticularis*.

Today, the grub of the palm weevil, *Rhynchophorus palmarum*, and the dynastine scarab *Oryctes* are favorite foods throughout the islands of the South Pacific. In French Polynesia these chitinous butterballs are prepared by being breaded in a combination of fine bread crumbs, spices, and fruit juice, then placed on a skewer and slowly roasted over charcoal. Other palm-infesting beetle grubs are consumed by various tribes in South America. The internal contents of the grubs are reported to have the consistency of mayonnaise and, when roasted, taste like beef marrow. Some tribes in South America encourage the standing crop of palm beetle grubs, creating new niches for the larvae by pulling down palms on purpose. Several months after the trees are felled, the trunks are inspected and the grubs harvested. Natives detect the presence of grubs by placing an ear against the trunk to listen for the grinding of the grubs' mandibles against the fibrous contents of the trunk.

Natives of Papua New Guinea harvest hundreds of pounds of palm grubs, rolling them up in banana leaves like giant sausages up to 3 meters in length. These grub rolls are cooked for a grub festival, and the stewed larvae are carefully distributed among members of the clan by the owner of the larvae. People locate the larvae by tapping fallen trees to detect the tunneling activities of the palm grubs, not unlike selecting a ripe watermelon at the market. In Vietnam palm grubs are dipped in sauce, deep-fried, wrapped in pastry, and refried.

Recipes for serving the larvae of melolonthine scarabs and cerambycids may be found in France and Japan and bear such scrumptious names as broiled or roasted beetle grubs, fried cockchafer grubs, roasted cockchafer grubs in paper, and cockchafer bouillon. Cockchafer bouillon is prepared and consumed to strengthen the nerves and was suggested as a special dish for those suffering from anemia. A recipe for this dish from an old French cookbook calls for one pound of cockchafers. After removing the elytra and legs, the prepared chafers are fried in two ounces of butter until crisp. Chicken stock is then added to the mixture, which is brought to a boil. After a small quantity of sliced calf liver is added, the mixture is ready to be served with chopped chives and croutons. Bon appetit!

Beetles are a conspicuous and relished component of modern Asian cuisine. In parts of Southeast Asia, larval and pupal buprestids and dytiscids are eaten. The Chinese collect large, adult dytiscids and hydrophilids under streetlights and prepare them in a number of ways. Basic preparation includes the removal of the elytra, wings, legs, and head. Water beetles may be fried in oil, seasoned, and eaten like nuts, or simply dropped in hot brine. These beetles are common in parts of the United States, but if your supply is low, you might be able to locate them in the shops of thriving Chinese communities throughout the larger cities of North America. According to the *Guinness Book of World Records*, the rarest condiment in the world is Cà Cuong, which is recovered in minute amounts from the secretions a chrysomelid beetle indigenous to Vietnam. During the height of the Vietnam War this coleopteran elixir sold for as much as $100 per ounce, but supplies dried up after 1975.

Thai peasants highly prize beetles as food, particularly the larvae, pupae, and adults of a particular dung scarab, which are collected from the excrement of their cattle and buffalo herds. Adult beetles are relieved of their elytra and other crunchy bits and are fried or roasted, sometimes in curries. The softer, more succulent larvae and pupae are soaked in coconut milk for several minutes before being roasted. This method results in a crispy exterior while rendering the interior into a consistency similar to that of a soufflé. The flavor is said to resemble a mixture of vegetables.

In the highlands of Mexico, cerambycid larvae are served with rice. Mexicans also brew a stimulating drink made from tiger beetles soaked in alcohol or water that is fermented and served. Australian Aborigines consider the large larvae of cerambycids an important part of their diet. The nut-flavored grubs are ripped from rotten logs and roasted like marshmallows over an open fire. In the United States, the mystique of eating the worm from a bottle of the Mexican liquor mescal has resulted in the marketing of a tequila-flavored lollipop with an entombed larva of the meal worm, *Tenebrio molitor*, a tenebrionid.

In Africa, larvae of the Goliath beetle are regarded as a treat. Collected at the base of banana trees, these larvae may reach nearly 15 centimeters in length. The Pedi of South Africa eat a variety of beetles, including large weevils and a large buprestid. Weevil-infested grain or meal are prepared without concern.

Most Americans react to the idea of eating insects with disbelief and disgust. Yet the consumption of insects in the United States is officially sanctioned by the Food and Drug Administration. This government agency regularly publishes a list of the maximum acceptable levels of insect adults, immatures, eggs, feces, and fragments in food items sold to the public. These standards differ depending on the food and exist because it is impossible to entirely exclude insects or parts of insects in processing and handling food items for human consumption. Since beetles represent the majority of insects and many are important crop pests, we can be assured that they form part of our balanced diet.

Canadian officials have used the confused flour beetle as a "good food guide." The beetles are fed new varieties of cereals and their rate of development is carefully monitored. The relative developmental rate of the beetles is directly proportional to the nutritional value of the grain variety, thereby allowing researchers to test the nutritional value of new hybrid grains early in their development. For the latest information on edible and delectable beetles and other insects, we highly recommend subscribing to *The Food Insects Newsletter*, which is edited by Dr. Florence Dunkel and is available through the Department of Entomology at Montana State University in Bozeman.

{TAKE TWO BEETLES & CALL ME IN THE MORNING}

THE real or imagined medicinal benefits of beetles probably developed as a result of their original consumption as food. The ingestion of large, powerful horned beetles is thought, in some parts of the world, to impart the same robust qualities to the consumer. Any therapeutic effects are probably due entirely to the power of suggestion, rather than to a direct physiological reaction.

Scarabs are a conspicuous element of most environments inhabited by humans and in the past were often credited with curative powers. The Roman scholar Pliny suggested simply the act of tying the melolonthine scarab *Polyphylla fullo* between two lizards was a remedy for the four-day fever, a form of malaria. Simply cutting the beetle in half and tying each section to the arm was considered to be helpful. Others prescribed certain scarabs, or their relatives, or parts thereof, to be worn in an attempt to treat convulsions, earaches, and other illness. Artificial fevers were sometimes induced by concoctions made from beetles, such as the stag beetles, to treat mental disorders, convulsions, and gout.

In Europe, pulverized carabids, chrysomelids, weevils, and coccinellids were used to relieve toothaches. Ground ladybirds were sold at the apothecary as *Pulvis dentrifricius*, or tooth powder. Ladybirds were also recommended as a cure for measles and colic. Other cures included carrying an uneven number of weevil larvae from the head of a thistle in a pouch worn around the neck. Having an even number of larvae was thought to render the treatment useless. Other beetles, such as gyrinids and large horned dynastine scarabs were used to concoct preparations that were purported to increase the libido. Even today, beetle horns and the ashes of stag beetles continue to be used in the preparation of aphrodisiacs in Turkey, while the Javanese click beetle *Oxynopterus mucronatus* is considered to be the primary ingredient of a particularly sexually stimulating potion. In Japanese folk medicine, species from several families of beetles were used to treat everything from convulsions and cancer to hydrophobia (fear of water) and hemorrhoids. The basis for these and other remedies remain obscured by time and myth.

Blister beetles of the family Meloidae are so designated because of the blister-producing properties of their defensive body secretion, cantharidin. The Spanish fly, *Lytta vesicatoria*, and the oil beetle, *Meloe*, are the two European species of meloids on which most of the literature is based. When cantharidin, which was isolated in 1812, penetrates the skin, it produces strong superficial irritation and blistering within a few hours.

Modern and ancient literature is filled with references to the uses and abuses of cantharidin, including as a treatment for epilepsy, sterility, asthma, rabies, and lesions resulting from gonorrhea. Today, cantharidin is probably best known for its purported aphrodisiac qualities, which were first noted in the sixteenth century. [PLATE 82] It apparently has a highly stimulating effect on the urinary tract, but even in small quantities it can be highly toxic. In eighteenth-century France one could partake of *pastilles à la Richelieu* or *bonbons à la Marquis de Sade* as a love potion to increase sexual performance. Recent catalogues mention cantharidin for use in ointments and plasters, the preparation of tinctures in veterinary medicine, and an ingredient in some hair restorers. In Peru, cantharidin extracted from a species of the genus *Pseudomeloe* is used to remove warts. The warts are scraped open and covered with a pulp made from the beetle. A blister forms and, after several days, the wart is destroyed.

Our own experiences with cantharidin range from a slight burning sensation of the fingertips while collecting the beetles by hand, to large, juicy blisters across the eyelids from carelessly wiping sweat from the brow immediately after having handled meloids. A colleague, while collecting on a particularly warm night in Mexico, noticed thousands of meloids, probably members of the genus *Epicauta*, swarming about the light under which he was working. Unfortunately, several of the beetles fell down his collar and rolled down his back, leaving numerous, yet intermittent, trails of burning, bubbling blisters. Some staphylinid beetles are also known to produce blistering in this manner.

Continued contact with beetles, willfully or otherwise, may lead to a hypersensitive reaction. Some people are highly allergic to the setae of dermestid beetles, which may infest carpets and animal hides. Allergies to

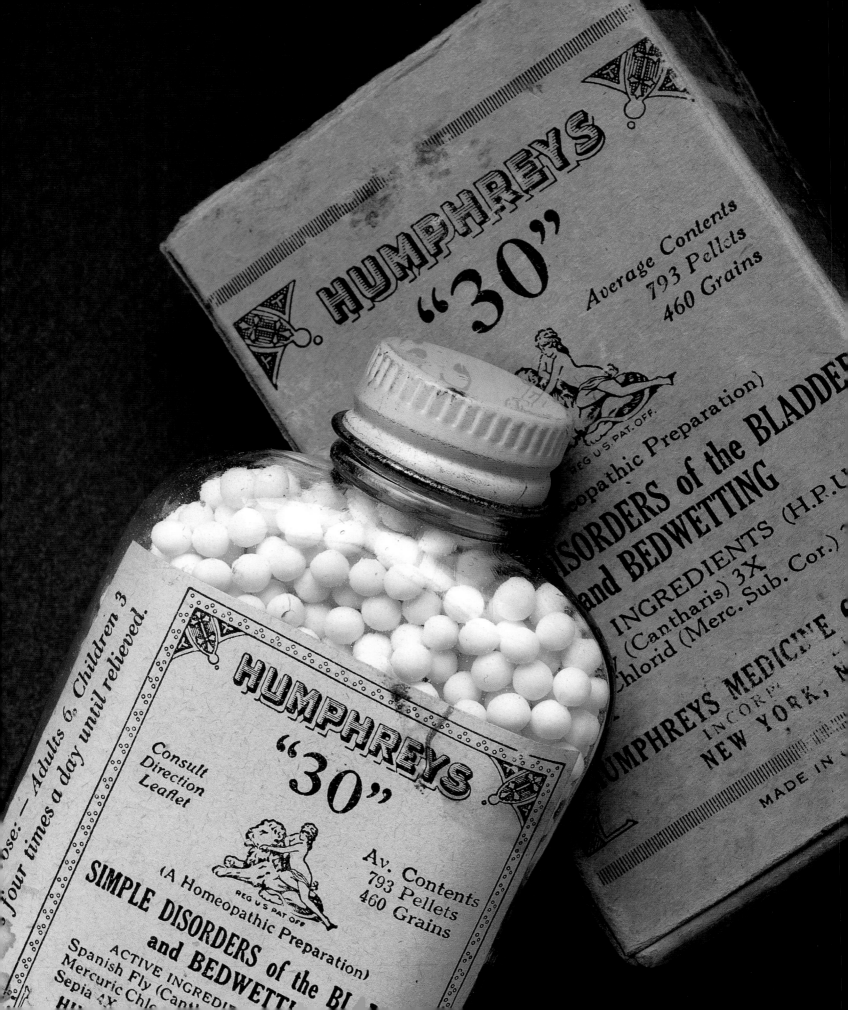

parts of arthropods are probably more common than we realize, since our exposure to these tiny fragments is frequent but rarely noticed. Heightened sensitivity to the shed skins of insects increases with prolonged exposure, as evidenced by those who come in contact with this material on a regular basis, particularly those working in insectaries (breeding facilities) and insect zoos.

Canthariasis—invasion of the human body by beetle larvae—is known but rare. Larvae of the churchyard beetle, *Blaps mortisaga*, have been known to infest the digestive tract, a result of superstitiously drinking foul graveyard water inhabited by the creatures. The ingestion of stored-grain pests, such as the meal worm, *Tenebrio molitor*, is also known to cause infestation. Scarabiasis is a rare condition, recorded in India, in which dung beetles of the genus *Onthophagus*, presumably in an effort to the reach the source of their favorite meal, become lodged within the rectum of sleeping children, causing discomfort and diarrhea. Removal of the beetle from this delicate area is best accomplished under medical supervision, since all the horns, spines, and other structures that facilitate the beetle's forward progression through soil and dung make its backward extrication difficult.

Not all medically important encounters with beetles are necessarily negative. Beetles with large, piercing mandibles have been used as crude sutures to close wounds. The insects are held up to the edge of the wound and are allowed to bite the skin, bringing the two sides of the open wound together. The heads of the insects are then separated from their bodies and left in place until the wound has healed.

Researchers at NASA's Goddard Space Flight Center discovered a unique use for the chemical reaction responsible for the luminescence of fireflies. Bacteria living in the human urinary tract that may cause severe kidney infections produce energy-rich ATP (adenosine triphosphate), which in the presence of luciferin and luciferinase, causes a light-producing reaction (see Chapter 4). Because the amount of light indicates the level of bacterial activity, the extent of the infection can be determined.

The use of beetles in folk medicine, often dismissed as myth and superstition, may prove to have a rational biomedical basis. Each species of beetle possesses a unique biochemical fingerprint that has been shaped, modified, and tested through millions of years of evolution and represents a virtually untapped wealth of biochemical information. E. O. Wilson states, "Biodiversity is our most valuable but least appreciated resource," and suggests that "a rare beetle sitting on an orchid in a remote valley of the Andes might secrete a substance that cures pancreatic cancer."

{THE INTERFACE BETWEEN BEETLES & HUMANS}

THE assessment of beetle qualities as "good" or "bad"—even the names that we give to beetles—merely reflect our own value system. Although we have touched on some of the more practical aspects of beetles in our lives, this can hardly be considered an accurate portrait of beetles and their role in the natural world. Beetles enrich our language, art, agriculture—even our diets—but what do we do for them? The fact is, we need beetles for our intellectual, spiritual, and biological survival, but they have very little use for us. Our challenge lies in our ability to recognize beetles as biological treasures, preserving them for future generations of, among others, etymologists, artisans, agriculturists, gourmands, and healers.

[PLATE 82]

The literature is filled with references to the uses and abuses of cantharidin, including as a treatment for epilepsy, sterility, asthma, rabies, lesions as a result of gonorrhea, bed wetting, and bladder disorders. Today, it is probably best known for its purported aphrodisiac qualities. *Courtesy of the Natural History Museum of Los Angeles County.*

~~~

six

BEETLEPHILIA

*The scientist does not study nature because it is useful. He
studies it because he delights in it and he delights
in it because it is beautiful.*

•

Jules Henri Poincare (1845-1912)
Quoted in *The Quest for Life in Amber,*
George and Roberta Poinar

COLEOPTERISTS have long been aware of the need to "get back to nature" in order to rejuvenate their minds, bodies, and research. Recently, scientists have begun to seriously study the inherent basis of the human need for nature, coined *biophilia* by E. O. Wilson. Wilson defines **biophilia** as the innately emotional affiliation of human beings to other living things. The biophilia hypothesis recognizes that our relationship with nature is complex and should be viewed not only from the perspective of what nature means to our immediate individual and collective survival, but also as a biologically based need essential to our emotional and intellectual development. Biophilia is the totality of our love/hate relationship with nature. Within this context we begin to explore the value of beetles.

{THE SIGNIFICANCE OF BEETLES}

THE biological basis for our affiliation with beetles may be viewed from the perspective of nine fundamental values that reveal how humans relate to nature: utilitarian, ecologistic, scientific, naturalistic, aesthetic, symbolic, dominionistic, negatavistic, humanistic, moralistic. These categories were developed by Stephen Kellert of Yale University in the late 1970s, not to define any instinctive behavior, but to identify clusters of learning behavior to understand basic human perceptions of animals. The elegance of this classification is that it may be considered a universal framework for the expression of basic human affinities for and dependence on the natural world. Applying this typology to our relationship with beetles, we can briefly summarize some of the ways in which these marvelous animals are meaningful in our lives.

[PLATE 83]

Many common, large brightly colored or horned beetles could be farmed for the dead-stock trade as a sustainable rain forest crop, generating in part, a local economy that encourages the preservation of the primary and secondary forests known to be the natural habitats of these beetles. Species of the genus *Plusiotis* could be raised in this manner. Left to right, top to bottom: *Plusiotis gloriosa*, USA; *Plusiotis victorina*, Mexico; *Plusiotis quetzalcoatli*, Mexico; *Plusiotis chrysargyrea*, Costa Rica; *Plusiotis resplendens*, Costa Rica; and *Plusiotis beyeri*, USA. *Specimens courtesy of the Natural History Museum of Los Angeles County and Paul McGray.*

If we accept our biophilic condition, our need for nature, then we must view all of nature as utilitarian. We have already discovered the nutritional, medicinal, and/or pest control value of some beetles. What other services might beetles provide us? The vast quantity of genetic and chemical information hidden in each beetle, detailing the successes and failures of millions of years of evolution, is virtually untapped. By exploiting beetles as a resource, we may discover new drugs or biological controls. By carefully recording the environmental preferences of existing beetles, we may gain additional insights into environments of the past and clues to what is in store for our planet in the future.

Beetles may also offer some motivation for preserving shrinking ecosystems. Our conservation strategy, particularly in the rain forests, must continue to shift from the idealistic preservation of "pristine" habitats to the realistic factoring in of the human component. If local peoples can derive an immediate financial return through a sustainable resource within the environment, they are more likely to preserve that environment. Beetles represent one such resource. Dead-stock trade in insects, mostly beetles and butterflies, runs into the tens of millions of dollars annually. [PLATE 83] Most of these specimens are decorative, used in the manufacture of ornaments and nonscientific displays. A small, but significant portion of these traded insects are sought by hobbyists, serious collectors, and bona fide researchers, who often deal in specimens of high value and quality. [PLATES 84 AND 85]

Butterfly farms, established in Central and South America, Malaysia, and Papua New Guinea, have been regarded as a benefit to butterfly conservation because they do not rely on specimens caught in the wild to supply commercial demand. Butterfly farmers build enclosed breeding sites to protect their stock from parasites and predators. When the adults emerge, a percentage of the stock population is released into local habitats to enhance the breeding potential of the wild population; the remaining specimens are packaged for sale. Proceeds from the sale of these butterflies support families, maintain farms, and enhance the surrounding environment.

Many common, large, brightly colored or horned beetles (buprestids, [PLATES 86, 87, AND 88] cerambycids, and ruteline, dynastine, and cetoniine scarabs) could also be farmed for dead-stock trade as a sustainable rain forest crop, generating, in part, a local economy that encourages the preservation of the remaining tracts of primary and secondary growth forest known to be the natural habitats of these beetles. However, the biological requirements—such as length of the larval stage, and subterranean or otherwise secretive habits—of many beetles make breeding efforts difficult and time-consuming.

The same ethical issues raised with more traditional animal stock could arise with beetle farming, particularly since they would be grown and slaughtered simply for decorative use and, as such, would steadily decrease in value. Beetles raised for research or to stock educational exhibits to raise the awareness of distant human populations about the role of insects in sustainable economies and to promote habitat preservation might offset these ethical concerns.

Beetles effectively function in the environment as two entirely different animals: larva and adult. Further, species that have even minor morphological differences may have distinctly different biologies, physiologies, and behaviors. Each species of beetle might easily provide a lifetime of observation and study; the depth and breadth of their scientific and ecological value is enormous. Their myriad forms and functions have helped to sharpen our current approaches to systematics of all life forms and to define our understanding of symbiosis, mimicry, and adaptation.

The presence or absence of certain beetle species in some environments may indicate even the most subtle of environmental perturbations. For example, some aquatic beetles can be used as **bioindicators** (a measure of the health of an ecosystem) in wetland management, since changes in water temperature favor the colonization of some species, while destroying others. Tiger beetles and other carabids have been suggested as bioindicators because of their stable taxonomy, small number of species, well-known biology, easily observed adults, and specialized habitat preferences. Other species of well-known beetles with specific habitat, host, and soil requirements, such as dung scarabs, buprestids, cerambycids, chrysomelids, curculionids, and tenebrionids might also serve as bioindicators in environmental studies.

Systematics research is an essential ingredient to the understanding of all biodiversity. It provides the text for evolution, reconstructing origins and relationships, and suggests how all organisms have adapted within dynamic environments. For example, a thorough knowledge of a closely related group, or clade, of beetles can lead, through inference, to reasonably accurate predictions of the unknown biologies of related taxa. On a broader scale, beetle systematics could provide information essential for assessing, designing, implementing, and ensuring the efficacy of environmental management practices and should be central to national environmental policies. The database that drives systematics research comes from specimen collections. Therefore, the human skills required to manage and identify these collections are a key resource for interpreting biodiversity.

THE SATISFACTION DERIVED FROM BEETLES

The naturalistic tendencies of many humans are reflected in the pleasure they derive from studying beetles and other arthropods in their natural environments. But few people understand why entomologists, amateur or professional, do what they do, particularly when the problems under investigation seem to involve no immediate economic gratification. Although many people are curious about the insects that they come across in their daily lives, few have had the opportunities or encouragement to explore the lives of these incredible creatures. Why?

Many would-be insectophiles were probably exposed to entomology via the ordeal of learning seemingly meaningless and unpronounceable names of scores of dead impaled or pickled organisms. Those of us who were fortunate enough to grow up in an insect-rich environment, or who had the opportunity to be exposed to entomology through a knowledgeable and enthusiastic mentor, were encouraged to go beyond the pervasive fears

[PLATE 84 AND 85]

The length of the larval stage and the secretive habits of some
beetles, such as the wood boring cerambycids may make
efforts to breed them difficult and time consuming. Plate 84
from top to bottom: *Rosenbergia straussi*, Papua New Guinea;
Rosenbergia weiski, Papua New Guinea. Plate 85 from top to
bottom: *Aristobia approximator*, Burma; and *Aristobia* sp.,
Thailand. *Specimens courtesy of Paul McGray.*

[PLATES 86, 87, AND 88]

Many common, large brightly colored or horned beetles could be farmed for the dead-stock trade as a sustainable crop, generating a local economy that encourages the conservation of the natural habitat. Plate 86: a coastal Namaqualand brush beetle, *Julodis chevrolati*, on the foliage of its host plant. Plate 87: *Themognatha mitchelli* is a large flower visiting buprestid found in Australia. Plate 88: *Evides pubiventris* is a buprestid beetle found thoughout eastern South Africa. *Photographs courtesy of Charles Bellamy.*

and misconceptions about beetles and other arthropods held by many people and perpetuated by the media. These lucky few share the realization that the study of beetles, or any other living thing, may reveal something unique about the world in which we live.

Entomology is one of the few biological disciplines in which anyone with a keen pair of eyes and an inquisitive mind can make valuable contributions to our understanding of the taxonomy, biology, and distribution of a highly significant life form. [PLATE 89] On a personal note, we, the authors, have always felt fortunate with our choice of career. The fact that we are paid to study these fascinating creatures outdoors in exotic, pristine locales only adds to our overall euphoria. In this regard we are professionals, but we will always remain amateurs in the truest sense of the word, which comes from the Latin *amare*, "to love."

Some coleopterists refer to their passion for beetles in terms of joy, excitement, wonder, delight, thrill, satisfaction, and fulfillment. The thrill of discovery is ever-present, since very little is known about the lives beetles lead. Even the most mundane of observations can led to new and exciting avenues of research. Einstein observed that study, being in general the pursuit of truth and beauty, is a sphere of activity in which we are permitted to remain children all of our lives.

THE AESTHETIC BEETLE

Beetles have inspired artisans for centuries. Their actual bodies have even been incorporated into pieces of jewelry, tapestries, and other decorative ornaments. People have celebrated the beauty of beetles by placing their colorful images on postage stamps around the world. The first insect stamp was a honeybee, issued in Nicaragua in 1891. Other countries quickly followed suit with silkworms, dragonflies, and sacred scarabs. By 1990, more than 300 countries and their protectorates had issued more than 1700 arthropod stamps, including 269 species of beetles, primarily scarabs, cerambycids, buprestids, curculionids, and carabids.

But it is not just the form of beetles that we find pleasing. For centuries we have been entranced by intricate behaviors that seem to mold beetles perfectly into innumerable environments: the subtle elegance of a chrysomelid escaping notice from its predators by resembling caterpillar frass; the gentle, rhythmic flashes of fireflies used to seduce their mates on a warm summer evening; the industriousness of dung beetles carving up the spoils of juicy bovine leavings, carting them away to provide for their young; the intricate wood carving of bark beetles. These are but a few examples of the glimpses we have had of the beauty of beetles as they interact with the environment.

THE SYMBOLIC BEETLE

Beetles have been used extensively in some cultures in metaphorical expression, often representing the unknown or unseen forces of nature (see Chapter 5). They have also appeared as the instruments of good and bad luck. They have influenced our language and expressive thought, contributing to the names of rock bands, automobiles, or derogatory terms such as *beetle-browed* or *grubby*. Our knowledge of beetles as symbols is increasing as more coleopterists delve into the cultural anthropological literature to ferret out intriguing examples from around the world.

MASTERY OVER BEETLES Our desire to control the world, or at least our immediate environment, is a very powerful drive—a drive that early in our history no doubt enhanced our chances of survival. Today, these tendencies are often associated with destruction and waste. But as Stephen Kellert points out, this perspective may be too narrow and is a result of exaggeration. After all, beetles are one of our greatest competitors for food, which explains in part why we spend millions of dollars annually in attempts to control the ravages of Japanese beetles, boll weevils, Colorado potato beetles, bark beetles, and a host of other species that infest products in storage, from massive grain silos to our home pantries.

Is it really necessary, or practical, to channel our rapidly dwindling resources to exterminate pest populations, when simple control to lower populations to levels that are not economically devastating (that is, **economic thresholds**) will suffice? Some of our excessive dominionistic tendencies seem to have clouded our ability to forge a more inclusive relationship with beetles, at least in terms of recognizing their positive role in a healthy environment.

THE FEAR OF BEETLES Fear is an essential component of human behavior and has played a significant role in shaping our survival skills. Within our more "civilized" environments, however, fear tends to surface more as a result of purposeful participation in dangerous sporting activities, helping to put an edge on our experiences as we push the envelope of our intellectual and physical limits. Although no specific fear of beetles has been defined, for many people beetles fall into the category of "creepy crawlies," contributing to an overall fear of insects, known as **entomophobia**. Despite their clear benefit, beetles and their spineless kin continue to be viewed with fear, antipathy, and aversion. In part, this response is an expression of our innate defense mechanism developed from having evolved together with these creatures, as well as our contemporary standards of cleanliness.

The sheer numbers of insects may profoundly threaten our notions of individuality and independence. Being so outnumbered by these armored legions presents a serious challenge to our sense of identity and superiority. This fear is exacerbated by their "alien" appearance, with modes of life and body forms unlike anything familiar, operating without apparent emotion or fear. The sad result of these irrational fears, biologically based as they may be, is that many of us are predisposed toward overreaction, a "zero tolerance" of beetles. Had we lacked this negativistic outlook on beetles, perhaps we would not have formulated and used such copious quantities of lethal insecticides in an attempt to eradicate them entirely.

FOR THE LOVE OF BEETLES As humans, we are instinctively drawn to animals that are closer to our own realm of experience, animals that are feathered or furry. We anthropomorphize (attribute human qualities to) many of the behaviors and responses of these animals, thereby establishing an emotional bond with them, far more easily than we ever could with beetles, which view the world with unblinking eyes and are encased in skinless, polished skeletons. Often characterized as unthinking and unfeeling, beetles rarely illicit from us the feelings of sympathy we easily afford cute and cuddly vertebrates. Even coleopterists, at least the less staid ones, may say that they "love" beetles, but this gush of unabashed emotion generally refers to the overall concept and pursuit of what is beetle, rather than an attachment to an individual insect.

CONSERVING BEETLES

We have briefly presented a framework for beetlephilia, our deep and necessary relationship with the Coleoptera, to demonstrate the significant contributions they make to our lives, beneficial and otherwise. By exploring their impact on our lives through a set of universally recognized values, we can begin to appreciate the practical uses of beetles as a natural resource. On an individual level, humans develop intellectually through the study of beetles. As a whole, society can begin to understand that beetles demonstrate principles that govern the lives of many other organisms. The metallic colors, arresting patterns, and unusual body forms of beetles have inspired generations of artisans and craftspeople. The fact that many beetles compete directly with us for food and shelter has resulted in a significant allocation of our resources directed at efforts to control them, with sometimes disastrous effects on the environment. Although they are one of the most numerous forms of life on the planet and have probably been so for millions of years, we still relate to them as alien invaders.

The lack of public awareness about beetles, not to mention all other invertebrates, is directly attributed to the simple fact that they are small and do not inspire the same feelings of sympathy that are elicited by the charismatic megafauna to which we seem to be inherently drawn. This lack of emotional connection with beetles often hampers our ability to develop a sense of ethical or moral responsibility toward beetles as expressed by efforts to conserve them. Nevertheless, as we begin to appreciate the essential role of beetles in our environment, their inclusion in environmental management plans seems inevitable.

{LEGISLATION TO PROTECT BEETLES}

THE Xerces Society, named after the extinct blue butterfly whose existence was erased from the San Francisco Bay area, and the Endangered Species Act (ESA) passed in 1973 have helped galvanize efforts to conserve invertebrates in the United States. The ESA requires the federal government to recognize and list beetles, among other organisms, that are threatened or endangered with extinction. As of 1994, nine species of beetles were afforded protection by the ESA: the American burying beetle, *Nicrophorus americanus*; the coffin cave beetle, *Batrigodes texanus*; the delta green ground beetle, *Elaphrus viridis*; Hungerfords crawling water beetle, *Brychius hungerfordi*; the Kretchmarr Cave mold beetle, *Texamaurops reddelli*; the northeastern beach tiger beetle, *Cicindela dorsalis dorsalis*; the puritan tiger beetle, *Cicindela puritana*; the Tooth Cave ground beetle, *Rhadine persephone*; and the Valley elderberry beetle, *Desmocerus californicus dimorphus*. [PLATE 90] These species are protected from any attempt to "harass, harm, pursue, shoot, wound, kill, trap, capture, collect, or attempt to engage in any such conduct." Any action that would modify or destroy the habitat of a listed species is considered a violation of the ESA.

In addition to the ESA, the Lacey Act and the Convention on International Trade in Endangered Species of Wild Fauna and Flora (CITES) are

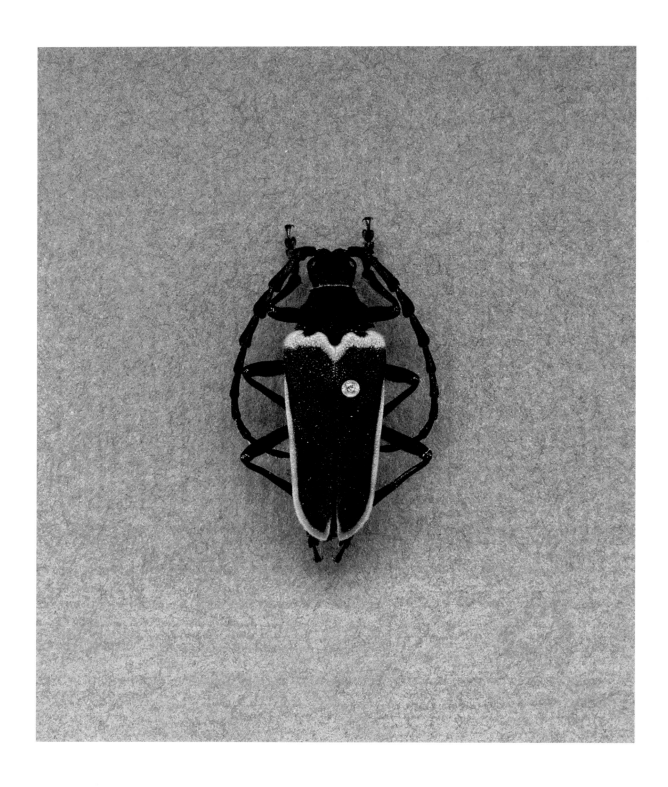

instrumental in protecting beetles in the United States and elsewhere. The original Lacey Act, passed in 1900, along with subsequent amendments, prohibits the interstate commerce of illegally killed wildlife and prevents the importation of certain types of wildlife thought to be injurious to agriculture. In 1981 the Lacey Act was amended with specific language that left no doubt that beetles, among other arthropods, are included under its protective powers.

Under the Lacey Act, any action undertaken to import, export, transport, sell, receive, or purchase, in interstate or foreign commerce, any beetle collected in violation of any federal, state, Native American tribal, or foreign law is a violation of federal law. For example, in a foreign country that requires a permit, collecting beetles without permission is a violation not only of that country's wildlife laws, but also of the Lacey Act. Collecting beetles without written permission in state or national parks is also prohibited by the Lacey Act. In addition, any infraction of the ESA automatically involves a violation of the Lacey Act. Falsifying label data or identifications of beetles with the intent to move them across state lines or international boundaries is also a federal offense under the Lacey Act.

CITES regulates the trade between signatory nations of species threatened with extinction. The current listing of species protected by CITES has three categories: (1) endangered species that are thought to be threatened by trade, (2) species that are not now threatened with extinction but may become so unless trade is strictly regulated, and (3) all species that are subject to local jurisdiction. The latter listing is provided to prevent or restrict exploitation by trade, which requires the cooperation of other signatory nations. As of 1992, no species of beetle was listed by CITES, despite several attempts to add species that are thought by some to be at risk through commercial exploitation by the dead-stock trade.

{LIVING ON THE EDGE}

THE Endangered Species Act includes provisions that encourage the acquisition of habitat, the development and implementation of recovery and habitat conservation plans, the negotiation of conservation agreements with local agencies, and the establishment and management of local preserves. All federal agencies use their authority to protect endangered beetles, but the U.S. Fish and Wildlife Service administers the ESA. Of all the beetles protected by the ESA, the American burying beetle, *Nicrophorus americanus*, is probably one of the best documented. [PLATE 91]

A recovery plan for the American burying beetle, designed by the Rhode Island division of the U.S. Fish and Wildlife Service, is part of an effort to increase the range and population of this beetle and secure its future. In addition, a protocol for rearing American burying beetles in the laboratory has been developed, detailing techniques such as how to determine the sex of beetles, specific materials needed to encourage successful copulation, and methods of establishing broods and sequestering adults. This protocol enables universities and zoos to breed these animals in captivity for subsequent release in the wild.

The American burying beetle is the largest carrion beetle in the United States, measuring 25 to 35 millimeters in length. It is a member of the family Silphidae, the burying, or sexton, beetles. In the not so distant past, every church had a sexton, or church caretaker, who provided a number of

[PLATE 90]

The valley elderberry beetle, *Desmocerus californicus dimorphus*, is one of nine species of beetles officially listed as endangered by the U.S. Fish and Wildlife Service. It is threatened by habitat loss. *Specimen courtesy of the Natural History Museum of Los Angeles County.*

[PLATE 91]

Of the nine endangered beetle species protected by the Endangered Species Act, *Nicrophorus americanus* is one of the best known. A recovery plan has been implemented to increase the population and range of this species to secure its future. *Photograph by Brett Ratcliffe.*

services, from ringing church bells to digging graves. The latter task is what came to mind as the carcass-burying activities of the larger, more spectacular silphids became known. American burying beetles are excellent subjects for investigations of behavior, sociobiology, and coevolution, because they exhibit one of the highest degrees of parental care known in the Coleoptera. By burying and preparing carcasses of birds and small mammals, burying beetles not only play an important role in nutrient recycling; their activities may directly lead to the control of populations of pestiferous flies and ants that depend on exposed carcasses for food and egg-laying sites.

A study of the silphids of Illinois in 1975 revealed that the American burying beetle had not been collected in the state for more than thirty years. Several entomologists in the early 1980s published field observations, literature records, and collection surveys suggesting that the American burying beetle was becoming more localized in its distribution, confirming suspicions that it had disappeared from much of its original distribution, which was the historical range of the northeastern deciduous forests. As of 1988, American burying beetles were thought to exist in only two widely separated populations. Block Island, off the coast of Rhode Island, is host to an apparently stable population of these beetles and has been the site for several studies of their life history. The eastern Oklahoma habitats are probably more typical of the historical ecosystems utilized by burying beetles, consisting of grasslands and oak-hickory and bottomland forests.

In 1989, on the basis of its disappearance throughout much of its historical range, the American burying beetle was added to the list of endangered species by the U.S. Department of Interior. The population of the American burying beetle is thought to have been dramatically reduced by a combination of factors resulting from deforestation and the subsequent decline of suitable animal carcasses. Deforestation not only disrupts the immediate environment; it also causes a shift in the forest fauna, either directly affecting the quality and quantity of carrion available, or reducing the amount of carrion available through increased competition by introduced scavengers.

Since being listed as an endangered species, American burying beetles have been found in Arkansas and Nebraska. Intensive field surveys conducted in Nebraska in 1994 located more than 300 beetles. These surveys will continue, striving to determine why the beetles seem to thrive in this state by examining the interplay of complex environmental and biological factors to understand the suite of positive and negative factors that influence their populations.

A similar European species, *Nicrophorus germanicus*, has also become less abundant and increasingly localized. Fortunately, the biology of this species was carefully studied before it began to disappear, providing clues that could be incorporated in the management of the less well known *Nicrophorus americanus*. Another large species, *Nicrophorus concolor*, is common in mature, undisturbed climax forests in Japan. This species has no doubt benefited from the environmental awareness of the Japanese people, which led to the preservation of much of this forest.

THE critical problem facing all regions of the world is the lack of taxonomic knowledge of beetles, exacerbated by the decline of another endangered species, systematic coleopterists. Nevertheless, the International Union for Conservation of Nature and Natural Resources (IUCN) has compiled the invaluable "Red List of Threatened Animals," documenting the threatened fauna of the world. In 1993, the Red List catalogued 378 species of beetles, variously categorized as endangered, vulnerable, rare, or suspected to be one of the above, but for which sufficient data are lacking. The number of beetle species listed as endangered outnumbers all other groups of animals included on the list. The Red List complements the IUCN "Red Data Books," of which the IUCN Invertebrate Red Data Book lists six species of beetles variously as vulnerable, rare, or endangered. Each listing is accompanied by a brief description and notes on distribution, habitat, ecology, scientific interest, threats to survival, and a list of proposed and enacted conservation measures.

CONSERVING BEETLES IN EUROPE   The former Soviet Union, Finland, and Sweden have produced red-data books that list beetles. Several other Scandinavian countries are currently preparing lists. Several European countries have crafted legislation listing species to be protected primarily from collectors, even though collecting has not been determined to be a serious threat to most beetle populations. Today, legislators and environmentalists have begun to recognize the importance of scientific collecting as a way to support effective recovery plans. But considering the historical popularity of insect collecting in Europe, legislation likely will continue to be extended to more species of beetles to protect them from nonscientific collecting activities.

Throughout the European community, regional, national, and international conventions extend protection to many beetles, including species of *Carabus*, *Cerambyx*, *Chrysocarabus*, *Cicindela*, *Cybister*, *Dytiscus*, *Ergates*, *Lucanus*, *Oryctes*, *Osmoderma*, *Polyphylla*, and *Rosalia*, to name a few. Of these, *Rosalia alpina*, *Ergates faber*, *Lucanus cervus*, and *Osmoderma eremita* are specifically protected throughout much of Europe. [PLATE 92]

The conservation of beetles must go beyond statutes that simply restrict collecting. Effective legislation must be based on an understanding of the roles that beetles play in the environment and include provisions for habitat preservation and research. Two European organizations actively promote the conservation of beetles on the basis of their ecological roles. The Water Beetle Specialist Group, part of the Species Survival Commission of the IUCN, recognizes the importance of aquatic beetles as bioindicators in wetland management in Europe and Southeast Asia. The Saproxylic Invertebrates Project focuses on selected groups of invertebrates, that are dependent, during some part of their life cycle, on standing or fallen trees or wood-inhabiting fungi. Buprestids, cerambycids, elaterids, lucanids, rhysodids, and scarabs are considered to be of possible use as bioindicators of the saproxylic system. The project strives to conserve and/or reestablish saproxylic organisms by establishing forest reserves, conducting surveys, restricting removal of downed wood, and improving forest management practices, such as controlled burning and habitat development to increase the availability of dead wood.

[PLATE 92]

Habitat destruction is the primary threat to beetles, but many laws are intended only to restrict collecting, which has not been shown to pose a direct threat to most beetle populations. Throughout the European community, regional, national, and international conventions extend protection to many beetles, including the cerambycid *Rosalia alpina*. *Specimen courtesy of the Natural History Museum of Los Angeles County.*

Habitat preservation in South Africa is relatively good but focuses primarily on the savannas
as home to the charismatic megafauna. Provin-
cial ordinances prohibit the collection and/or trading and, in some cases,
restrict the export and/or transport of all species listed as endangered.
Insects within national parks and other specially designated wilderness
areas are also protected by law. Currently, only two genera of beetles are
afforded protection in South Africa; both of them come under the jurisdiction of the former Cape Province. [PLATE 93]

The first protected beetle, a wingless canthonine scarab, *Circellium bac-
chus*, is the largest member of its tribe in the world. Once widespread
throughout northern and eastern Transvaal, Zimbabwe, and Namibia, it
now appears to be restricted to the Cape Province, where it is protected by
law. In the Addo Elephant National Park, this beetle uses both elephant
and buffalo dung, but prefers the latter, to construct its brood balls. Since
management efforts within the national park have focused on larger animals, it was not immediately understood why the removal of buffalo from
the park to an area with better grazing would depress the beetle's already
slow rate of reproduction. This situation highlights the problem posed by
conserving animals that are on an entirely different scale within a single
reserve and provides a daunting challenge to conservation managers.

The second genus afforded protection is the stag beetle *Colophon*,
which has thirteen slow-moving, wingless species. [PLATE 94] *Colophon* is
restricted to the mountaintops of the Cape high-mountain habitat, with
many of the peaks or groups of peaks home to a unique species. The males
of some species are characterized by grotesquely formed mandibles and
forelegs. The entire genus was given provincial protection in April 1992,
partly as a result of the extremely high prices commanded by specimens
offered for sale in Europe, the United States, and Japan. Prices for complete
sets of *Colophon* have been advertised at upward of $10,000, and single spec-
imens of the rarest species have been listed at several thousand dollars.
Despite their legal protection, some commercial insect dealers continue to
collect, trade, and sell these rare beetles. A recent attempt to list *Colophon*
in the first category of the CITES endangered species list failed, but South
African authorities are planning to petition that *Colophon* be listed in the
third category, a measure that offers limited international protection.

SAVING BEETLES
DOWN UNDER Nationally, Australian beetles do not enjoy specific
legal protection, but they are protected under the
umbrella of all indigenous fauna. The Wildlife Pro-
tection Act of 1982 requires that specimens to be exported for commercial
use must have been bred in captivity or collected under an approved management program. Several species of lucanids and cetoniine scarabs are regularly exported from Australia through the specialized dead-stock trade.
The demand for specimens of some of the more beautiful species, such as
the lucanid *Phalacrognathus muelleri*, has raised concerns in the conservation
community because of the difficulty of rearing this species in captivity.

At the provincial level, Western Australia protects the entire family of
jewel beetles, the Buprestidae. This legislation was enacted primarily over
concern about the intensive efforts to collect these beetles in the region, not
because any of the species in question were threatened with extinction.

[PLATE 93]

Habitat protection in South Africa focuses primarily on the
savannas as home to the charismatic megafauna. However, all
wildlife residing in national parks and nature preserves are
afforded protection, including this small brush beetle,
Neojulodis picta, that lives in the Great Karoo. *Photograph by
Charles Bellamy.*

The combination of New Zealand's cool temperate climate with glaciation, volcanism, and tectonic movement has created numerous ecological niches favoring the evolution of many unusual beetles. Strong winds, stable habitats, and the lack of predators have encouraged the development of many large, wingless species. Upon arriving in the islands, humans destroyed some habitats as they milled the forests and burned the tussock grasslands. The introduction of exotic predators, pests, and diseases has also had a significant impact on the entire biota. The New Zealand Department of Conservation, charged with conserving New Zealand's beetle fauna, in 1992 officially recognized twenty-four species of beetles—mostly carabids, scarabs, lucanids, and curculionids—as endangered. Eventually, a recovery plan will be drawn up for each of these endangered beetles.

Rats are a particular threat to wingless species of weevils. The Campbell Island ribbed weevil, *Heterexis seticostatus*, is a small wingless weevil that as an adult inhabits the base of a native lily, *Bulbinella rossi*, and as a larva feeds on the roots. Even though the lily is becoming more common, the weevil population is threatened by an overabundance of the introduced Norway rat. The large flightless *Oclandius laeviusculus* weevil, on the other hand, lives in restricted sites in the Auckland Islands that are, fortunately, inaccessible to the ravages of feral pigs, which uproot the weevil's host plants.

The Cook Strait and Chatham Island amychus beetles, *Amychus granulatus* and *Amychus cendezei*, respectively, are primitive barklike click beetles that have lost the ability to click. Adults are usually found in rock crevices or beneath stones and logs, while the larvae are thought to feed on the roots of forest trees and shrubs. Despite much deforestation, amychus beetles were reported as abundant fifty years ago. Today they are considered rare, even though the vegetation has changed little during the past half century. The recovery of an endangered reptile, the tuatara, is suspected to be contributing to their decline.

The Cromwell chafer, *Prodontria lewisi*, is a wingless scarab found only in slightly disturbed stretches of shallow loamy sand at the junction of the Clutha and Kawarau rivers. But this habitat has been reduced to about a hundred hectares by the development of a hydroelectric project. Extensive surveys in surrounding areas have failed to locate additional populations of the beetle. While introduced rabbits are overgrazing much of this beetle's host plant, the beetles themselves are being devoured by another European immigrant, the little owl, introduced to New Zealand just after the turn of the twentieth century to control orchard birds. In 1981–1982 the Cromwell Chafer Scientific Reserve was set aside to study this threatened beetle.

At least two species of New Zealand beetles are presumed extinct: *Megacolabus sculpturatus*, a weevil known only from a single specimen collected at Akaroa, Banks Peninsula, more than a hundred years ago; and *Xyloteles costatus*, a cerambycid beetle from the Chatham Islands that has not been collected in more than fifty years.

[PLATE 94]

The lucanid *Colophon* is restricted to the mountaintops of the Cape region of South Africa. Despite their legal protection, some dealers continue to collect, trade, and sell these uncommon beetles. Shown here is *Colophon westwoodi*, which inhabits the top of Table Mountain near Cape Town. *Specimen courtesy of the Natural History Museum of Los Angeles County.*

HABITAT loss is the primary threat to the survival of all beetles. Despite their incredible success, beetle populations are limited, sometimes drastically so, by factors such as fire, urbanization, acid rain, electric lights (including "bug zappers"), overgrazing, agricultural expansion, water impoundment, pollution, deforestation, soil erosion, persistent adverse weather, off-road recreational vehicles, and logging, to name a few. [PLATE 95] The grazing of cattle on public lands has been shown to have disastrous effects on butterfly populations, yet we are absolutely ignorant as to the effects of grazing on soil-dwelling beetles. The effects of pesticide drift (the potential impact of pesticide moving into non-target habitats) and competition from exotic organisms must have an enormous impact on beetle populations, yet are poorly documented. Even the use of parasites and pathogens, promoted as environmentally friendly and economical, are self-dispersing and may produce unpredictable and irreversible effects on some beetle populations.

Conservationists, legislators, and government agencies are concerned about collecting as a threat to beetles. Beetle collectors may be viewed, with some validity, as megapredators, consuming all in their path with terrifying voracity, preying on individual species and destroying their habitats. In reality, however, beetle collectors are merely hazards to a relatively small number of individuals, and only rarely, when combined with the loss of habitat, does collecting pose a threat to an entire population of beetles. The annual harvest of vast numbers of big, showy beetles for dead-stock trade throughout the world surely has an impact on their populations, but to date, no evidence of beetles being driven to extinction by commercial exploitation has been presented.

Nevertheless, as conservationists, we must consider that the merciless commercial exploitation of a species, combined with widespread habitat loss caused by destructive logging and farming practices, may result in an appreciable decline of some beetle populations. The resultant genetic bottleneck could lead to a novel adaptation or to the evolutionary coup de grâce of extinction. Whether or not these collecting activities affect beetle populations, the question is, Is such wholesale slaughter of specimens caught in the wild purely for decorative purposes, regulated or not, ethical?

One of the most significant events affecting systematic entomology was the passage in 1981 of sweeping amendments to the Lacey Act that had been enacted by the U.S. Congress in 1900. In 1992, federal wildlife agents began searching insect collections, both institutional and personal, for specimens suspected to have been collected illegally under the Lacey Act. These federal investigations led to the conviction of three amateur butterfly collectors, who pled guilty to charges of selling and trading specimens collected illegally in Mexico and in state and national parks in the United States.

Beetle collection and systematic research has not been the same since, for the very laws designed to protect wildlife now appear to be a serious impediment to international beetle research. Institutions may hesitate to accept specimens of important scientific value, including type specimens, if they are not accompanied by the proper documentation, possibly resulting in the unnecessary loss of data. Further, museum curators overseas have

[PLATE 95]

Off-road recreational vehicles are a threat to beetles restricted to remote sand dunes, such as this North American ruteline scarab, *Pseudocotalpa giulianii*. Although not listed as endangered, this and many other sand loving species enjoy the limited protection afforded to all organisms living on lands under the jurisdiction of the Bureau of Land Managment. *Specimen courtesy of the Natural History Museum of Los Angeles County.*

expressed some reluctance to ship specimens to the United States for research, fearful that shipments opened for inspection by wildlife authorities may be confiscated, damaged, or lost.

Fortunately, the ultimate result of the initial confusion is bound to be positive, for it will ensure that legislators, conservationists, and coleopterists finally come together to craft regulations to conserve beetles on the basis of their own biology. Most laws were written to protect vertebrates and plants, whose reproductive capacities and mobility are no match for most insect species. Laws that encourage managed collecting, as well as reasonable procedures for issuing permits, will help ensure the free flow of data in the form of specimens, adding strength to the systematic and ecological foundation of our environmental management policies.

The fact remains that responsible collecting by professionals and amateurs results in collections that provide information on distribution, seasonality, and abundance—data essential to our efforts to conserve beetles. Amateur collections are particularly useful in filling gaps found in institutional collections. The potential use of beetle collections in environmental education is enormous, and many coleopterists attribute the beginnings of their scientific interest to collections of beautifully curated beetles. The very act of collecting provides essential training in understanding the biology of beetles. Any attempt to place unreasonable restrictions on collecting will have disastrous effects on the recruitment of future coleopterists. While legislation may be necessary for the recovery of rare species, it should not restrict the useful gathering of information by the nonspecialist. Entomological societies that have adopted voluntary codes of collecting ethics that urge restraint and encourage responsibility in the collecting of specimens of rare or localized species may provide the needed compromise.

{BEETLES AS TOOLS IN ENVIRONMENTAL EDUCATION}

THE nine values that we discussed earlier can be used as a template for a holistic biodiversity education program that seeks to develop a close kinship between humans and all the rest of nature. Such an education program would introduce concepts that emphasize the interrelationships of organisms, thereby establishing a greater understanding of the role of biodiversity in maintaining the quality of human life and creating a deep ethical concern for the environment. This ethic must embrace the knowledge that only through a rich and diverse biosphere will humans, collectively and individually, flourish physically, emotionally, and intellectually.

The familiar, yet bizarre, nature of beetles makes them the perfect ambassadors for environmental awareness. [PLATE 96] But winning public support for beetles and most other invertebrates requires a carefully executed educational program that incorporates all learning styles and environments. Museums and zoos have a central role in countering "vertebrate chauvinism" by developing awareness of invertebrates through exhibits, education, and research programs. The educational, inspirational, and entertainment value of a well-designed invertebrate exhibit can be significant. Live and preserved beetles on display, static exhibits, and outreach programs can effectively introduce the public in the incredible diversity of our biological heritage. The fascination generated by the color, form, and behavior of live beetles, supplemented with informative displays on their

[PLATE 96]

The familiar, yet bizarre nature of beetles makes them the
perfect ambassadors for environmental awareness. Here one
of the authors (Evans) shares the biology of the Central
American hercules beetle, *Dynastes hercules*, with young
museum visitors. *Photograph by Dick Meier, courtesy of the
Natural History Museum of Los Angeles County.*

diversity and interrelationships with humans, help visitors appreciate the importance of all invertebrates.

The need for the conservation of beetles can be highlighted in an insect zoo by the incorporation of captive breeding projects of endangered species. For example, the Invertebrate Conservation Centre (ICC) at the London Zoo and the Valbonne Insectarium in Italy have produced hundreds of individuals of the Olympia ground beetle, *Carabus olympiae*, a species listed as endangered by the IUCN, in an attempt to reintroduce them into the wild. [PLATE 97] During this project, the staff of the ICC not only gained valuable experience in the culture of this species; it also acquired information on the identification and treatment of their pathogens. Insect World at the Cincinnati Zoo has raised several generations of American burying beetles for release in the wild, and other zoos will soon follow suit. Increased public awareness of the plight of endangered beetles through programs like these could help raise funds to support other beetle recovery programs.

Live beetles can easily be incorporated in the classroom. They are readily available in urban environments, require little space, are inexpensive to maintain, and are easily displayed. Caring for beetles in the classroom instills a sense of responsibility among the students, helping to develop their awareness that beetles, just like humans, have basic environmental and nutritional needs that must be met, or they will perish. Simple experiments with beetles can show the effects of temperature on the rate of development, food preferences, and a variety of other interesting behaviors. One of the most effective, yet least understood, elements of beetle education is the power of touch. To talk about the defensive strategies of tenebrionid beetles is one thing, but placing a live desert ironclad beetle in a child's hand to demonstrate how it feigns death to escape the notice of predators is an experience that will not soon be forgotten!

{A MULTIPLICITY OF INTERRELATIONSHIPS}

FOR more than thirty years the media has bombarded us with a picture of the apocalypse in which an unprecedented loss of biological diversity has resulted from, in large part, the loss of habitats because of unrestricted human population growth, purposeful and/or accidental biological introductions, and overexploitation. We relate to this catastrophe as if it were a sporting event, except that the scores are expressed in units of the hundreds or thousands of plant and animal species on the brink of extinction. If present rates continue, the extinction rate in the next two centuries will exceed that of the Cretaceous-Tertiary boundary, which marked the disappearance of the dinosaurs. The vast majority of this loss will be invertebrates, including many beetles. The general public will not fully realize the impact of this loss until it is too late, because science simply lacks the ability to assess a disaster of such magnitude. The root cause of this preventable catastrophic event, unlike previously documented extinctions, will not be the subject of long debate by future civilizations, for the responsibility will be squarely on our shoulders.

Only our arrogance allows us to think that we can "save" the biological world. Nature is a set of continually changing processes, influenced by the interactions of physical and chemical forces and continually modulated by biological phenomena. It will continue without us. Humans have been pre-

[PLATE 97]

The Invertebrate Conservation Centre at the London Zoo has launched numerous captive breeding projects of threatened and endangered invertebrates, including the Olympia ground beetle, *Carabus olympiae*. The information gathered from this and other breeding efforts can be used to conserve other species of *Carabus* which may become threatened with extinction. Top, left to right: *Carabus auratus*, France; *Carabus coriaceous*, Germany; and *Carabus hortensis*, Czechoslovakia. Bottom, left to right: *Carabus superbus*, Germany; *Carabus aurenitens*, Germany; *Carabus* sp., France; and *Carabus insulicola*, Japan. *Specimens courtesy of the Natural History Museum of Los Angeles County.*

ceded by several mass extinctions, and we have no reason to think that we can change this pattern of events. Yet, the fact that we cannot control nature is no reason to ignore our collective impact on the environment, for Earth is all we have. We have nowhere else to go.

Our desire to preserve biodiversity has arisen solely because of the realization that we, as an evolving species, are in trouble. By preserving biodiversity, we acknowledge our interdependence with nature and enhance our chances to persist on life's stage a little longer. This realization manifests itself, in part, as beetlephilia. The value of preserving beetles, a significant portion of our biodiversity, depends on human perceptions and our economic, health, and agricultural concerns. [PLATE 98] If we must treat beetles as a commodity, then it is essential that we possess and use the knowledge that will enable us, as a collection of cultures, to appreciate both the short- and the long-term benefits of keeping beetles around. Through our efforts to preserve beetle diversity we will accumulate the "capital" of information to invest in our own future. The "stock reports" of beetles are gathered from a thorough knowledge of the systematic, taxonomic, and ecological data.

We should consider adopting E. O. Wilson's necessarily anthropocentric ethic when viewing the natural world. First, biodiversity is the Creation. In other words, it is all we have. Second, other species are our kin. All multicellular eukaryotic organisms—that is to say, all living things whose cells include a nucleus—from the simplest organisms to beetles to humans, apparently evolved from a single population of organisms nearly 2 billion years ago. Third, the biodiversity of each country is a part of its natural heritage, the biota and geography of which reflect a deep, national history that predates its citizens. Finally, biodiversity is the frontier of the future. The knowledge gained from its study flows through our science, art, and public affairs.

The presence or absence of beetles will stand as the measure of how we are doing in protecting the biosphere for present and future generations, for beetles play a significant part of a seasonal exchange between earth and sky, a pulse in the cycle of life. Each beetle is but part of a population and embodies the sum total of its evolutionary history and potential. Each population interacts with the others, including our own, and with the soil and atmosphere in a multiplicity of interrelationships that melt seamlessly into one another. We can take solace in beetlephilia.

[PLATE 98]

The value of preserving beetles, a significant portion of our biodiversity, depends upon human perceptions of our economic, health, and agricultural concerns. Shown here is the cetoniine scarab, *Agestrata semperi*, a male and female, from the Philippines. *Specimens courtesy of the Natural History Museum of Los Angeles County.*

[EPILOGUE]

NATURE, and thus the study of beetles, is important—fundamentally important—to us, the human species. Unlike the objects and devices we design and construct or the short-lived fads that wax and wane in popularity, nature is timeless. Beetles are not infinitely timeless, since all contemporary forms can be traced back to the at least hypothetical mother of all beetles, but they are nearly so. And even though things that are relatively timeless are not necessarily endless, we are far from discovering the end to beetle diversity.

Many people are oblivious to nature, some because their lives are filled with the simple pursuits, not unlike those of beetles, of feeding and caring for themselves and their offspring. Others are hedonistic, ignoring nature and her creatures in their pursuit of pleasure and effectively reducing the life of a beach or a desert sand dune, for example, to something that just gets in the way of a roaming dune buggy.

All aspects of nature are mysterious, perhaps even imposing, because of our lack of knowledge about them. We could think of exploring nature as a first visit to a vast academic library, with floors and floors of dusty books, filled with information—if only we knew where to start. But whether the mystery before us is a vast library or a group of beetles in a forest, the only way to increase understanding is through study—to open the books and learn.

No matter what habitat you find yourself in, you will find beetles. At first you might see them simply as beetles, endless variations of a common timeless theme, but as you continue to adjust the focus, you will see them as a mixture of distinct and individual species, each with a name, each becoming more familiar. Their names are keys to the compartments of an incredibly vast world of knowledge. Knowing the names will expand your understanding and will introduce you to their lives, their distribution and, through a knowledge of their relatives, to their past history. By developing our understanding on an individual basis, we develop more respect and love for the whole of nature. Then instead of rushing past or over the varied habitats and life forms of our world, we might occasionally stop to watch, listen, learn, appreciate, and then teach.

We are glad to have the opportunity to teach you something about beetles. Now we hope that you will want to watch and appreciate them. You never know when you might find one that has never been seen before.

"If and when the day comes when pure science is once again generally appreciated as a self-justifying intellectual adventure of mankind, then the coleopterists should be able to step forward and claim their share of its glory."
Roy A. Crowson, Glasgow
The Biology of the Coleoptera

BEETLE FAMILIES[1]

Order COLEOPTERA

•Suborder Archostemata
  1. Family Ommadidae
  2. Family Crowsoniellidae
  3. Family Micromalthidae (telephone pole beetles)[2]
  4. Family Cupedidae (reticulated beetles)
•Suborder Myxophaga
  5. Family Lepiceridae
  6. Family Torridincolidae
  7. Family Hydroscaphidae (skiff beetles)
  8. Family Microsporidae (minute bog beetles)
•Suborder Adephaga
  9. Family Gyrinidae (whirligig beetles)
  10. Family Haliplidae (crawling water beetles)
  11. Family Trachypachidae
  12. Family Noteridae (burrowing water beetles)
  13. Family Amphizoidae (trout-stream beetles)
  14. Family Hygrobiidae
  15. Family Dytiscidae (predaceous diving beetles)
  16. Family Rhysodidae (wrinkled bark beetles)
  17. Family Carabidae (ground beetles)

•Suborder Polyphaga
· Series Staphyliniformia
Superfamily Hydrophiloidea
  18. Family Hydrophilidae (water scavenger beetles)
  19. Family Sphaeritidae (false clown beetles)

  20. Family Synteliidae
  21. Family Histeridae (hister, or clown, beetles)

Superfamily Staphylinoidea
  22. Family Hydraenidae (minute moss beetles)
  23. Family Ptiliidae (feather-winged beetles)
  24. Family Agyrtidae
  25. Family Leiodidae (round fungus beetles)
  26. Family Scydmaenidae (antlike stone beetles)
  27. Family Silphidae (carrion beetles)
  28. Family Staphylinidae (rove beetles)

· Series Scarabaeiformia
Superfamily Scarabaeoidea
  29. Family Lucanidae (stag beetles)
  30. Family Passalidae (bessbugs)
  31. Family Trogidae (hide, or skin, beetles)
  32. Family Glaresidae
  33. Family Pleocomidae (rain beetles)
  34. Family Diphyllostomatidae
  35. Family Geotrupidae (earth-boring dung beetles)
  36. Family Belohinidae
  37. Family Ochodaeidae
  38. Family Ceratocanthidae
  39. Family Hybosoridae
  40. Family Glaphyridae
  41. Family Scarabaeidae (dung beetles, chafers)

· Series Elateriformia
Superfamily Scirtoidea
  42. Family Decliniidae
  43. Family Eucinetidae (plate-thigh beetles)
  44. Family Clambidae (minute beetles)
  45. Family Scirtidae

Superfamily Dascilloidea
  46. Family Dascillidae (soft-bodied plant beetles)
  47. Family Rhipiceridae (cedar beetles)

Superfamily Buprestoidea
  48. Family Schizopodidae (false buprestid beetles)
  49. Family Buprestidae (jewel, or metallic wood-boring, beetles)

Superfamily Byrrhoidea
  50. Family Byrrhidae (pill beetles)
  51. Family Elmidae (riffle beetles)

Superfamily Dryopoidea
  52. Family Dryopidae (long-toed water beetles)
  53. Family Lutrochidae
  54. Family Limnichidae (minute marsh-loving beetles)
  55. Family Heteroceridae (variegated mud-loving beetles)
  56. Family Psephenidae (water-penny beetles)
  57. Family Cneoglossidae
  58. Family Ptilodactylidae (toed-winged beetles)
  59. Family Chelonariidae
  60. Family Eulichadidae
  61. Family Callirhipidae

Superfamily Elateroidea
  62. Family Artematopidae
  63. Family Brachypsectridae (Texas beetles)
  64. Family Cerophytidae
  65. Family Eucnemidae (false click beetles)
  66. Family Throscidae
  67. Family Elateridae (click beetles)
  68. Family Plastoceridae
  69. Family Drilidae
  70. Family Omalisidae
  71. Family Lycidae (net-winged beetles)
  72. Family Telegeusidae (long-lipped beetles)
  73. Family Phengodidae (glow-worms)
  74. Family Lampyridae (lightning bugs or fireflies)
  75. Family Omethidae
  76. Family Cantharidae (soldier beetles)
  77. Family Podabrocephalidae
  78. Family Rhinorhipidae

[1] Modified from Lawrence and Newton (1995). Lawrence, J.F. and A.F. Newton, Jr. 1995. "Families and subfamilies of Coleoptera (with selected genra, notes, references and data on family group names)." In Pakaluk, J. and S.A. Slipinski, eds., *Biology, Phylogeny, and Classification of Coleoptera*. Papers Celebrating the 80th Birthday of Roy A. Crowson. Muzeum I Instytut Zoologii PAN, Warsaw, pp. 779-1006.
[2] Common names are given only for large families.

· Series Bostrichiformia
Superfamily Derodontoidea
79. Family Derodontidae (tooth-necked fungus beetles)

Superfamily Bostrichoidea
80. Family Nosodendridae (wounded-tree beetles)
81. Family Dermestidae (skin beetles)
82. Family Bostrichidae (powder-post beetles)
83. Family Anobiidae (death-watch beetles)
84. Family Jacobsoniidae

· Series Cucujiformia
Superfamily Lymexyloidea
85. Family Lymexylidae (ship-timber beetles)

Superfamily Cleroidea
86. Family Phloiophilidae
87. Family Trogositidae (bark-gnawing beetles)
88. Family Chaetosomatidae
89. Family Cleridae (checkered beetles)
90. Family Acanthocnemidae
91. Family Phycosecidae
92. Family Prionoceridae
93. Family Melyridae (soft-winged flower beetles)

Superfamily Cucujoidea
94. Family Protocucujidae
95. Family Sphindidae (dry-fungus beetles)
96. Family Brachypteridae
97. Family Nitidulidae (sap beetles)
98. Family Smicripidae
99. Family Monotomidae
100. Family Boganiidae
101. Family Helodidae
102. Family Phloeostichidae
103. Family Silvanidae
104. Family Passandridae
105. Family Cucujidae (flat bark beetles)
106. Family Laemophloeidae
107. Family Propalticidae
108. Family Phalacridae (shining flower beetles)
109. Family Hobartiidae
110. Family Cavognathidae
111. Family Cryptophagidae (silken fungus beetles)
112. Family Lamingtoniidae
113. Family Languriidae (lizard beetles)
114. Family Erotylidae (pleasing fungus beetles)
115. Family Byturidae (fruitworm beetles)
116. Family Biphyllidae (false skin beetles)
117. Family Bothrideridae
118. Family Cerylonidae
119. Family Alexiidae
120. Family Discolomidae
121. Family Endomychidae (handsome fungus beetles)
122. Family Coccinelidae (ladybird beetles)
123. Family Corylophidae (minute fungus beetles)
124. Family Lathridiidae (minute brown scavenger beetles)

Superfamily Tenebrionoidea
125. Family Mycetophagidae (hairy fungus beetles)
126. Family Archeocrypticidae
127. Family Pterogeniidae
128. Family Ciidae (minute tree-fungus beetles)
129. Family Tetratomidae
130. Family Melandryidae (false darkling beetles)
131. Family Mordellidae (tumbling flower beetles)
132. Family Rhipiphoridae (wedge-shaped beetles)
133. Family Colydiidae (cylindrical bark beetles)
134. Family Monommidae
135. Family Zopheridae
136. Family Ulodidae
137. Family Perimylopidae
138. Family Chalcodryidae
139. Family Trachelostenidae
140. Family Tenebrionidae (darkling beetles)
141. Family Prostomidae
142. Family Synchroidae
143. Family Oedemeridae (false blister beetles)
144. Family Stenotrachelidae
145. Family Meloidae (blister beetles)
146. Family Mycteridae
147. Family Boridae
148. Family Trictenotomidae
149. Family Pythidae
150. Family Pyrochroidae (fire-colored beetles)
151. Family Salpingidae (narrow-waisted bark beetles)
152. Family Anthicidae (antlike flower beetles)
153. Family Aderidae (antlike leaf beetles)
154. Family Scraptiidae

Superfamily Chrysomeloidea
155. Family Cerambycidae (long-horned beetles)
156. Family Megalopodidae
157. Family Orsodacnidae
158. Family Chrysomelidae (leaf beetles)

Superfamily Curculionoidea
159. Family Nemonychidae
160. Family Anthribidae (fungus weevils)
161. Family Belidae (primitive weevils)
162. Family Attelabidae (leaf-rolling weevils)
163. Family Brenthidae
164. Family Caridae
165. Family Ithyceridae
166. Family Curculionidae (weevils, or snout beetles)

MAJOR WORLD BEETLE COLLECTIONS

ARGENTINA

Museo Argentino de Ciencias Naturales
Division Entomologia
Av. Angel Gallardo 470
1405 Buenos Aires

Museo de La Plata
Division Entomologia
Universidad Nacional de La Plata
Paseo del Bosque
1900 La Plata

AUSTRALIA

Australian National Insect Collection
Division of Entomology
C.S.I.R.O.
P.O. Box 1700
Canberra ACT 2601

Australian Museum
P.O. Box A285
Sydney NSW 2000

Macleay Insect Collection
Macleay Museum
University of Sydney
Sydney NSW 2006

Queensland Museum
P.O. Box 3300
South Brisbane QLD 4101

South Australian Museum
North Terrace
Adelaide SA 5000

Museum of Victoria
Department of Entomology
Abbotsford VIC 3067

Western Australia Museum
Spider and Insect Collection
Francis Street
Perth WA 6000

AUSTRIA

Naturhistorisches Museum Wien
Postfach 417
Burgring 7
1040 Wien

BELGIUM

Institut Royal des Sciences Naturelles de
Belgique
Collections Nationales Belges d'Insectes et
d'Arachnides
29, Rue Vautier
B-1040 Brussels

Musee Royal de l'Afrique Centrale
Section d'Entomologie
Leuvensesleenweg 13
B-3040 Tervuren

BRAZIL

Instituto Nacional de Pesquisas da Amazìo-
nia
Colecáo Sistemática da Entomologia (INPA)
Estrada do Aleixo, 1756
C.P. 478, 69.011 Manaus, Amazonas

Museu de Entomologia Pe. Jesus Santiago
Moure
Universidade Federal do Paraná
Departamento de Zoologia
C.P. 19020, 81531-970 Curitiba, Paraná

Museu Nacional
Quinta da Boa Vista
São Cristovao
20.942 Rio de Janeiro, RJ

Museu de Zoologia da Universidade de São
Paulo
Biblioteca, 7172
01.051 São Paulo, SP

BULGARIA

Insect Collection
National Museum of Natural History
Bulgarian Academy of Sciences
Boulv. Tsar Osvobodital
BG-1000 Sofia

Insect Collection
Institute of Zoology
Bulgarian Academy of Sciences
Boulv. Tsar 1
BG-1000 Sofia

CANADA

Strickland Entomological Museum
Department of Entomology
University of Alberta
Edmonton, Alberta T6G 2E3

Canadian National Collection of Insects
Centre for Land and Biological Resources
Research
Biological Research Division, Agriculture
Canada
Ottawa, Ontario K1A 0C6

Canadian Museum of Nature
Entomology Division
P.O. 3443 Station D
Ottawa, Ontario K1P 6P4

CHILE

Colección Nacional de Insectos
Museo Nacional de Historia Natural
Casilla 787, Santiago

CHINA

Insect Collection
Institute of Zoology
Academia Sinica
19 Zhongguancun Lu
Haidan 100080 Beijing

Museum of Shanghai
Institute of Entomology
Academia Sinica
225 Chungking Road (S.)
Shanghai 200025

COLOMBIA

Colección Entomologica "Luis Maria
Murillo"
ICA, Tibaitatá, Apto. AÇreo 151123
Eldorado, Bogotá

COSTA RICA

Instituto Nacional de Biodiversidad
Apto. 22-3100
Santo Domingo de Heredia
3100 Heredia

CROATIA

Entomology Collection
Hrvatski Narodni Zoološki Muzej
Demetrova Ul. 1
4100 Zagreb

CUBA

Museo Nacional de Historia Natural
Capitolio Nacional
La Habana 2
Ciudad de La Habana 10200

CZECH REPUBLIC

Department of Entomology
National Museum of Natural History
Kunratice 1
148 00 Praha 4

DENMARK

Department of Entomology
Zoological Museum
University of Copenhagen
Universitetsparken
DK-2100 København

ECUADOR

Quito Catholic Zoology Museum
Departamento de Biologia
Pontificia Universidad Catolica del Ecuador
12 de Octubre y Carrion
Apto. 2184 Quito

EGYPT

Insect Collection
Department of Entomology
Faculty of Science
Cairo University
Giza, Cairo

FINLAND

Zoological Museum
Finnish Museum of Natural History
University of Helsinki
P. Rautatiek 13
SF-00100 Helsinki

FRANCE

National Collection of Insects*
Musée National d'Histoire Naturelle
45, Rue Buffon
Paris 75005

* The largest beetle collection in the world.

GERMANY

Museum für Naturkunde der Humboldt Universität zu Berlin
Bereich Zoologisches Museum
Invalidenstrasse 43
D-1040 Berlin

Zoologisches Forschungsinstitut und Museum "Alexander Koenig"
Adenaueralle 160
D-5300 Bonn 1
Deutsches Entomologisches Institut
Eberswalde Finow 1, D-1300

Zoologische Staatssammlung
Munchhausenstrasse 21
D-8000 München 60, Bayern

HUNGARY

Zoological Department
Hungarian Natural History Museum
Baroos Utca 13
H-1088 Budapest

INDIA

National Zoological Collection
Zoological Survey of India
34, Chittarangjan Avenue
Calcutta 700 012

ISRAEL

Insect Collection
Zoological Museum
Tel Aviv University
Tel Aviv 69978

ITALY

Museo Zoologico "La Specola"
Via Romana 17
50125 Firenze

Museo Civico di Storia Naturale "Giacomo Doria"
Via Brigata Liguria 9
I-16121 Genoa

Spinola Collection
Museo Regionale Scienze Naturali
Via Giolitti 36
Torino 10123

JAPAN

Entomological Collections
National Science Museum (Natural History)
Hya Kunin-Cho 3-23-1
Shinjuku-ku, Tokyo 160

KENYA

Section of Entomology
National Museums of Kenya
P.O. Box 40658
Nairobi

MALAYSIA

Natural History Division
Sarawak Museum
Kuching 93566

Sarawak
National Museum
Damansara Road
Kuala Lumpur

MEXICO

Colección Entomologica
Instituto de Biologia
Universidad Nacional Autonomá de Mexico
Apdo. Postal 70133
04510 Mexico D.F.

NAMIBIA

State Museum of Namibia
P.O. Box 1203
Windhoek 9000

NETHERLANDS

Insituut voor Taxonomische Zoologie
Zoologisch Museum
Universitet van Amsterdam
Plantage Middenlaan 64
1018 DH Amsterdam

Nationaal Natuurhistorische Museum
Raamsteeg 2
Leiden 2311 Pl.

NEW ZEALAND

New Zealand Arthropod Collection
Landcare Research New Zealand Ltd.
Private Bag 92170
Auckland

NICARAGUA

Servicio Entomológico Autónomo
Museo Entomológico
SEA, Apartado Postal 527
León

PANAMA

Smithsonian Tropical Research Institute
P.O. Box 2072
Balboa

PERU

Colección del Departmento de Entomologia
Museo de Historia Natural
Universidad Nacional Mayor de San Marcos
Av. Arenales 1267, Apartado 14-0434
Lima 14

POLAND

Museum of the Institute of Zoology
Polish Academy of Science
Wilcza 64
00-679 Warszawa

RUSSIA

Zoological Institute
Russian Academy of Science
St. Petersburg

SOUTH AFRICA

Entomology Department
South African Museum
P.O. Box 61
Cape Town 8000

Coleoptera Department
Transvaal Museum
P.O. Box 413
Pretoria 0001

South African National Collection of Insects
Private Bag X134
Pretoria 0001

SPAIN

Museo Nacional de Ciencias Naturales
Paseo de La Castellana 84
Madrid

SWEDEN

Naturhistoriska Riksmuseet
Sektionen fur Entomologi
S-10405 Stockholm

Zoological Museum
Uppsala University
P.O. Box 561
S-75122 Uppsala

SWITZERLAND

Entomology Department
Naturhistorisches Museum
Augustinergasse 2
CH-4001 Basel

Muséum d'Histoire Naturelle
Case Postale 434
CH-1211 Geneva

UNITED KINGDOM

The Natural History Museum[1]
Department of Entomology
Cromwell Road
London SW7 5BD

Linnean Society[2]
Burlington House
Piccadilly
London WIV OLQ

Hope Entomological Collections
University Museum
Parks Road
Oxford OXI 3PW

UNITED STATES

National Museum of Natural History[3]
Smithsonian Institution
Washington, DC 20560

Department of Entomology
American Museum of Natural History
Central Park West at 79th St.
New York, NY 10024

Entomology Section
Natural History Museum of Los Angeles
County
900 Exposition Blvd.
Los Angeles, CA 90007

J. Linsley Gressitt Center for Research
in Entomology
Department of Entomology
Bernice P. Bishop Museum
P.O. Box 19000A
Honolulu, HI 96819

Insect Collection
Field Museum of Natual History
Roosevelt Road and Lake Shore Drive
Chicago, IL 60605

Entomology Department[4]
Museum of Comparative Zoology
Harvard University
26 Oxford Street
Cambridge, MA 02138

Department of Entomology
California Academy of Sciences
Golden Gate Park
San Francisco, CA 94118

URUGUAY

Colección Entomologica del Museo Nacional
de Historia Natural
Casilla de Correo 399
Montevideo

ZIMBABWE

Invertebrate Collection
National Museum
P.O. Box 240
Centenary Park
Bulawayo

[1] One of the most important collections in the world.
[2] The collection of Linnaeus.
[3] The largest beetle collection in the western hemisphere.
[4] The collections of LeConte and Horn.

Appendix 3

PROFESSIONAL SOCIETIES
DEDICATED TO BEETLE STUDY

Unlike the worldwide Lepidoptera Society and its many regional chapters, no global societies are yet dedicated to the study of beetles. In fact, only five societies that we know of focus their attention on Coleoptera.

AUSTRIA

Wiener Coleopterologenverein
A-1010 WIEN (Vienna)
Burgring 7
• Founded in 1912. Publication: *Koleopterologische Rundschau.*

JAPAN

The Japan Coleopterological Society
Osaka
• Founded in 1945. Publication: *Entomological Review of Japan.*

The Japanese Society of Coleopterology
Tokyo
• Founded in 1973. Publication: *Elytra.*

SPAIN

Association Europea de Coleopterologia
Departament de Biologia Animal - Artrïpodes
Facultat de Biologia
Universitat de Barcelona
Avda. Diagonal, 645
08028 Barcelona
• Founded in 1987. Publications: *Elytron*, an annual journal; *Elytron*, *Supplement*, an irregular journal; *Coleopterological Monographs*, an irregular series for larger papers; *Advances in Coleopterology*, irregular proceedings of international congresses of coleopterology.

UNITED STATES

The Coleopterists Society
c/o Ed Zuccaro
23 D'Evereaux St.
Natchez, MS 39120
• Founded in 1947. Publication: *The Coleopterists Bulletin*, a quarterly journal.

SELECTED REFERENCES

Chapter 1

Arnett, R. H. 1971. *The Beetles of the United States*. American Entomological Institute, Ann Arbor, Mich.

Blackwelder, R. E. 1957. "Checklist of the coleopterous insects of Mexico, Central America, the West Indies, and South America." *United States National Museum Bulletin* 185(6):927–1492.

Blackwelder, R. E., and R. M. Blackwelder. 1948. *The Leng Catalogue of Coleoptera of America, North of Mexico. Fifth supplement, 1939–1947*. John D. Sherman, Jr., Mount Vernon, N.Y.

Cotterill, F. P. D. 1995. "Systematics, biological knowledge and environmental conservation." *Biodiversity and Conservation* 4:183–205.

Crowson, R. A. 1981. *The Biology of the Coleoptera*. Academic Press, London.

Dupuis, C. 1984. "Willi Hennig's impact on taxonomic thought." *Annual Review of Entomology* 15:1–24.

Erwin, T. L. 1982. "Tropical forests: Their richness in Coleoptera and other arthropod species." *Coleopterists Bulletin* 36(1):74–75.

Erwin, T. L. 1991. "How many species are there?: Revisited." *Conservation Biology* 5(3):330–333.

Erwin, T. L., and J. C. Scott. 1980. "Seasonal and size patterns, trophic structure, and richness of Coleoptera in the tropical arboreal ecosystem: The fauna of the tree *Luehea seemannii* Triana and Planch in the Canal Zone of Panama." *Coleopterists Bulletin* 34(3):305–322.

Gaston, K. J. 1991. "The magnitude of global insect species richness." *Conservation Biology* 5(3):283–296.

Mayr, E. 1988. *Toward a New Philosophy of Biology. Observations of an Evolutionist*. Belknap Press of Harvard University, Cambridge, Mass.

Pakaluk, J., and S. A. Slipinski, eds. 1995. *Biology, Phylogeny, and Classification of Coleoptera: Papers Celebrating the 80th Birthday of Roy A. Crowson*, 2 vols. Muzeum i Instytut Zoologii PAN, Warsaw.

Stork, N. E. 1988. "Insect diversity: Fact, fiction and speculation." *Biological Journal of the Linnaean Society* 35:321–337.

Strong, D. R., J. H. Lawton, and R. Southwood. 1984. *Insects on Plants. Community Patterns and Mechanisms*. Harvard University Press, Cambridge, Mass.

White, R. E. 1983. *Field Guide to the Beetles of North America*. Peterson Field Guide Series. Houghton Mifflin, Boston.

Chapter 2

Arrow, G. J. 1951. *Horned Beetles. A Study of the Fantastic in Nature*. Dr. W. Junk, The Hague.

Bailey, W. J. 1991. *Acoustic Behavior of Insects. An Evolutionary Perspective*. Chapman and Hall, London.

Berenbaum, M. R. 1989. *Ninety-nine Gnats, Nits, and Nibblers*. University of Illinois Press, Urbana.

Booth, R.G., M.L. Cox, and R.B. Madge. 1990. *IIE Guides to Insects of Importance to Man. 3. Coleoptera*. International Institute of Entomology, Natural History Museum, London.

Borror, D. J, C.A. Triplehorn, and N.F. Johnson, eds. 1989. *An Introduction to the Study of Insects*. Saunders, Philadelphia.

Brown, L., and L. L. Rockwood. 1986. "On the dilemma of horns." *Natural History* 104(7):55–61.

Caveney, S. 1986. "The phylogenetic significance of ommatidium structure in the compound eyes of polyphagan beetles." *Canadian Journal of Zoology* 64(9):1787–1819.

Chapman, R. F. 1971. *The Insects. Structure and Function*. Elsevier, New York.

Crawford, C. S. 1981. *Biology of Desert Invertebrates*. Springer-Verlag, Berlin.

Crowson, R. A. 1981. *The Biology of the Coleoptera*. Academic Press, London.

Darwin, C. 1871. *The Descent of Man, and Selection in Relation to Sex*, vol. 1. D. Appleton, New York.

Eberhard, W. G. 1980. "Horned beetles." *Scientific American* 242(3):166–182.

Hadley, N. F. 1993. "Beetles make their own waxy sunblock." *Natural History* 102(8):44–45.

Janzen, D. H. 1983. "*Pelidnota punctulata* (Comecocornizuelo, Ant-acacia beetle)." In D. H. Janzen, ed., *Costa Rican Natural History*. University of Chicago Press, Chicago, pp. 753–754.

Kistner, D. H. 1982. "The social insect's bestiary." In *Social Insects*, vol. 3. Academic Press, New York, pp. 1–244.

Kistner, D. H. 1990. "The integration of foreign insects into termite societies or why do termites tolerate foreign insects in their societies?" *Sociobiology* 17(1):191–215.

Lawrence, J. F., and E. B. Britton. 1994. *Australian Beetles*. Melbourne University Press, Carleton, Australia.

Matthews, R. W., and J. R. Matthews. 1978. *Insect Behavior*. Wiley, New York.

Preston-Mafham, R., and K. Preston-Mafham. 1993. *The Encyclopedia of Land Invertebrate Behaviour*. MIT Press, Cambridge, Mass.

Richards, A. G. 1951. *The Integument of Arthropods. The Chemical Components and Their Properties. The Anatomy and Development. The Permeability*. University of Minnesota Press, Minneapolis.

Thornhill, R., and J. Alcock. 1983. *The Evolution of Insect Mating Systems*. Harvard University Press, Cambridge, Mass.

Snodgrass, R. E. 1935. *Principles of insect morphology*. McGraw Hill, New York.

Stehr, F. W. 1991. *Immature Insects*, vol. 2. Kendall/Hunt, Dubuque, Iowa.

Thornhill, R., and J. Alcock. 1983. *The Evolution of Insect Mating Systems*. Harvard University Press, Cambridge, Mass.

Wigglesworth, V. B. 1950. *The Principles of Insect Physiology*. Methuen, London.

Chapter 3

Barr, T. C., Jr., and J. R. Holsinger. 1985. "Speciation in cave faunas." *Annual Review of Ecology and Systematics* 16:313–337.

Carlson, R. W., and F. B. Knight. 1969. "Biology, taxonomy, and evolution of four sympatric *Agrilus* beetles (Coleoptera: Buprestidae)." *Contributions of the American Entomological Institute* 4(3):1–105.

Coope, G. R. 1979. "Late Cenozoic fossil Coleoptera: evolution, biogeography, and ecology." *Annual Review of Ecological Systematics* 10:247–267.

Elias, S. A. 1994. *Quaternary Insects and Their Environments*. Smithsonian Institution Press, Washington, D.C.

Hespenheide, H. A. 1991. "Bionomics of leaf-mining insects." *Annual Review of Entomology* 36:535–560.

Howarth, F. G. 1983. "Ecology of cave arthropods." *Annual Review of Entomology* 28:365–389.

Illies, J. 1983. "Changing concepts in biogeography." *Annual Review of Entomology* 28:391–406.

Mani, M. S. 1962. *Introduction to High Altitude Entomology*. Methuen, London.

Poinar, G. O. 1992. *Life in Amber*. Stanford University Press, Stanford, Calif.

Poinar, G. O., and R. Poinar. 1994. *The Quest for Life in Amber*. Addison-Wesley, Reading, Mass.

Williams, G. A. 1993. *Hidden Rainforests: Subtropical Rainforests and Their Invertebrate Diversity*. New South Wales University Press, Kensington, Australia.

Chapter 4

Alcock, J. 1994. "Postinsemination associations between males and females in insects: The mate-guarding hypothesis." *Annual Review of Entomology* 39:1–21.

Alpert, G. D. 1994. "A comparative study of the symbiotic relationships between beetles of the genus *Cremastocheilus* (Coleoptera: Scarabaeidae) and their host ants (Hymenoptera: Formicidae)." *Sociobiology* 25(1):1–276.

Evans, D. L., and J. O. Schmidt, eds. 1990. *Insect Defenses. Adaptive Mechanisms and Strategies of Prey and Predators*. State University of New York Press, Albany.

Gressitt, J. L. 1966. "Epizoic symbiosis: The Papuan weevil genus *Gymnopholus* (Leptopiinae) symbiotic with cryptogamic plants, oribatid mites, rotifers and nematodes." *Pacific Insects* 8(1):221–280.

Gressitt, J. L. 1977. "Symbiosis runs wild in a Lilliputian 'forest' growing on the back of high-living weevils in New Guinea." *Smithsonian* 7(11):135–140.

Hoagland, M., and B. Dodson. 1995. *The Way Life Works*. Times Books, Random House, New York.

Holldobler, B. 1971. "Communication between ants and their guests." *Scientific American* 224(3):86–93.

Holm, E., and J. F. Kirsten. 1979. "Pre-adaptation and speed mimicry among Namib Desert scarabaeids with orange elytra." *Journal of Arid Environments* 2:263–271.

Kistner, D. H. 1982. "The social insect's bestiary." In H. R. Hermann, ed., *Social Insects*, vol. 3. Belknap Press of Harvard University Press, Cambridge, Mass., pp. 23–45.

Lloyd, J. E. 1984. "The occurence of agressive mimicry in fireflies." *Florida Entomologist* 67:368–374.

Matthews, E. G. 1972. "A revision of the scarabaeine dung beetles of Australia. I. Tribe Onthophagini." *Australian Journal of Zoology*, Suppl. Series 9.

Matthews, R. W., and J. R. Matthews. 1978. *Insect Behavior*. Wiley, New York.

McIver, J. D., and G. Stonedahl. 1993. "Myrmecomorphy: Morphological and behavioral mimicry of ants." *Annual Review of Entomology* 38:351–379.

Milne, L. J., and M. Milne. 1976. "The social behavior of burying beetles." *Scientific American* 235(2):84–89.

Murlis, J. 1992. "Odor plumes and how insects use them." *Annual Review of Entomology* 37:505–532.

Rettenmeyer, C. W. 1970. "Insect mimicry." *Annual Review of Entomology* 15:43–74.

Reyes-Castillo, P., and M. Jarman. 1980. "Some notes on larval stridulation in Neotropical Passalidae (Coleoptera: Lamellicornia)." *Coleopterists Bulletin* 34(3): 263–270.

Shapiro, A. M., and A. H. Porter. 1989. "The lock-and-key hypothesis: Evolutionary and biosystematic interpretation of insect genitalia." *Annual Review of Entomology* 34:231–245.

Timm, R. M., and J. S. Ashe. 1988. "The mystery of the gracious hosts." *Natural History* 97(9):6–10.

Vander Meer, R. K., and D. P. Wojcik. 1982. "Chemical mimicry in the myrmecophilous beetle *Myrmecaphodius excavaticollis*." *Science* 218:806–808.

Wilson, E. O. 1971. *The Insect Societies*. Belknap Press of Harvard University Press, Cambridge, Mass.

Zeh, J. A., and D. W. Zeh. 1994. "Tropical liaisons on a beetle's back." *Natural History* 103(3):36–42.

Chapter 5

Adams, J., ed. 1992. *Insect Potpourri: Adventures in Entomology*. Sandhill Crane Press, Gainesville, Fla.

Arnett, R. H., and R. L. Jacques. 1985. *Insect Life. A Field Entomology Manual for the Amateur Naturalist*. Prentice-Hall, Englewood Cliffs, N.J.

Berenbaum, M. R. 1995. *Bugs in the System*. Helix Books, Addison-Wesley, Reading, Mass.

Brown, R. W. 1956. *Composition of Scientific Words. A Manual of Methods and a Lexicon of Materials for the Practice of Logotechnics*. Roland W. Brown, Baltimore.

Cambefort, Y. 1994. "Beetles as a religious symbol." *Cultural Entomology Digest* No. 2:15–21.

Davies, M., and J. Kathirithamby. 1986. *Greek Insects*. Gerald Duckworth, London.

Defoliart, G. 1989. "The human use of insects as food and as animal feed." *Bulletin of the Entomological Society of* America 35(2):23–35.

Desmond, A., and J. Moore. 1992. *Darwin*. Penguin Books, London.

Essig, E. O. 1931. *A History of Entomology*. Macmillan, New York.

Evans, G. 1975. *The Life of Beetles*. Hafner, New York.

Hagen, K. S., and J. M. Franz. 1973. "A history of biological control." In R. F. Smith, T.E. Mittler, C.N. Smith, eds., *History of Entomology*. Annual Reviews, Inc., Palo Alto, Calif., pp. 433–476.

James, M. T., and R. F. Harwood. 1969. *Herm's Medical Entomology*. Macmillan, New York.

Hogue, C. L. 1993. *Latin American Insects and Entomology*. University of California Press, Berkeley.

Hubbell, S. 1993. *Broadsides from Other Orders. A Book of Bugs*. Random House, New York.

Klausnitzer, B. 1981. *Beetles*. Exeter, New York.

Newman, L. H. 1966. *Man and Insects*. Natural History Press, Garden City, N.Y.

Ratcliffe, B. C. 1990. "The significance of scarab beetles in the ethnoentomology of non-industrial indigenous peoples." In *Ethnobiology: Implications and Applications*. Proceedings of the First International Congress of Ethnobiology (Museu Paraense Emilio Goeldi, Belém, Brazil). pp. 159–181.

Rivers, V. Z. 1994. "An overview of beetle elytra in textiles and ornaments." *Cultural Entomology Digest* No. 2:2–9.

Schwabe, C. W. 1979. *Unmentionable Cuisine*. University Press of Virginia, Charlottesville.

Sharrell, R. 1971. *New Zealand Insects and Their Story*. Collins Brothers, Auckland, New Zealand.

Sutton, M. Q. 1988. *Insects as Food: Aboriginal Entomophagy in the Great Basin*. Ballena Press Anthropological Papers, No. 33. Ballena Press, Menlo Park, California.

Taylor, R. L. 1975. *Butterflies in My Stomach. Or: Insects in Human Nutrition*. Woodbridge Press, Santa Barbara, Calif.

Waterhouse, D. F. 1974. "The biological control of dung." *Scientific American* 230:100–109.

Chapter 6

Anderson, R. S. 1982. "On the decreasing abundance of *Nicrophorus americanus* Olivier (Coleoptera: Silphidae) in eastern North America." *Coleopterists Bulletin* 36(2):362–365.

Collins, N. M. 1987. "Legislation to conserve insects in Europe." AES Pamphlet No. 13. Amateur Entomologists' Society, London.

Collins, N. M., and J. A. Thomas, eds. 1991. *The Conservation of Insects and Their Habitats*. 15th Symposium of the Royal Entomological Society of London, 14–15 September 1989. Academic Press, London.

Cotterill, F. P. D. 1995. "Systematics, biological knowledge and environmental conservation." *Biodiversity and Conservation* 4:183–205.

Creighton, J. C., C. C. Vaughn, and B. R. Chapman. 1993. "Habitat preference of the endangered American burying beetle (*Nicrophorus americanus*) in Oklahoma." *Southwestern Naturalist* 38(3):275–306.

Davis, L. R., Jr. 1980. "Notes on the beetle distributions, with a discussion of *Nicrophorus americanus* Olivier and its abundance in collections (Coleoptera: Scarabaeidae, Lampyridae, Silphidae)." *Coleopterists Bulletin* 34(2):245–251.

Endrody-Younga, S. 1988. "Evidence for the low-altitude origin of the Cape mountain biome derived from the systematic revision of the genus *Colophon* Gray (Coleoptera: Lucanidae)." *Annals of the South African Museum* 96(9):359–424.

Hamel, D. R. 1991. *Atlas of Insects on Stamps of the World*. Tico Press, Falls Church, Va.

Hamilton, G., and G. R. Phelps. 1992. "Zoos and conservation education." *International Zoo Yearbook* 31:97–98.

Kellert, S. R. 1993. "Values and perceptions of invertebrates." *Conservation Biology* 7(4):845–855.

Kellert, S. R. 1993. "The biological basis for human values of nature," pp. 42-69. In S. R. Kellert and E. O. Wilson, eds., *The Biophilia Hypothesis*. Island Press, Washington D.C.

Kellert S. R. 1996. *The Value of Life. Biological Diversity and Human Society*. Island Press, Washington D.C.

Kozol, A. J. 1992. "A guide to rearing the American burying beetle, *Nicrophorus americanus*, in captivity." Unpublished report to the U.S. Fish and Wildlife Service.

Lomolino, M. V., J. C. Creighton, G. D. Schnell, and D. L. Certain. 1995. "Ecology and conservation of the endangered American burying beetle (*Nicrophorus americanus*)." *Conservation Biology* 9(3):605–614.

Malausa, J. C., and J. Drescher. 1991. "The project to rescue the Italian ground beetle." *International Zoo Yearbook* 30:75–79.

Meads, M. 1990. *Forgotten Fauna. The Rare, Endangered, and Protected Invertebrates of New Zealand*. Department of Scientific and Industrial Research, Wellington, New Zealand.

Meehan, C. 1995. "Education: Improving the image of invertebrates." *Victorian Naturalist* 112(1):60–62.

Mendoza, J. G. 1995. "The entomologist and wildlife laws." *American Entomologist* Summer:75–76.

New, T. R. 1994. "Invertebrate interests in the World Conservation Union's Species Survival Commission." *Memoirs of the Queensland Museum* 36(1):147–151.

Opler, P. A. 1991. "North American problems and perspectives in insect conservation." In N. M. Collins and J. A. Thomas, *The Conservation of Insects and Their Habitats*. 15th symposium of the Royal Entomological Society of London, 14-15- September 1989. Academic Press, London, pp. 9–32.

Pearce-Kelly, P. 1994. "Invertebrate propagation and re-establishment programmes: The conservation and education potential for zoos and related institutions." In P. J. S. Olney, G. M. Mace, and A. T. C. Festner, *Creative Conservation: Interactive Management of Wild and Captive Animals*. Chapman & Hall, London, pp. 329–337.

Pearce-Kelly, P., D. Clarke, M. Robertson, and C. Andrews. 1991. "The display, culture and conservation of invertebrates at London Zoo." *International Zoo Yearbook* 30:21–30.

Pearson, D. L., and F. Cassola. 1992. "World-wide species richness of tiger beetles (Coleoptera: Cicindelidae): Indicator taxon for biodiversity studies." *Conservation Biology* 6(3):376–391.

Pyle, R., M. Bentzien, and P. Opler. 1981. "Insect conservation." *Annual Review of Entomology* 26:233–258.

Ratcliffe, B. C., and M. L. Jameson. 1992. "New Nebraska occurrences of the endangered American burying beetle (Coleoptera: Silphidae)." *Coleopterists Bulletin* 46(4):421–425.

Robinson, M. H. 1993. "Invertebrates: The key to holistic education." In *Invertebrates in Captivity Conference, SASI/ITAG Conference, Tucson, Arizona, August 13-15, 1993*. pp. 1–11. Sonoran Arthropod Studies Institute, Tucson, Arizona.

Sagan, D., and L. Margulis. 1993. "God, Gaia, and biodiversity." In S. R. Kellert and E. O. Wilson, eds., *The Biophilia Hypothesis*. Island Press, Washington D.C., pp. 345–364.

Samways, M. J. 1994. *Insect Conservation Biology*. Chapman & Hall, London.

Scholtz, C. H., and S. L. Chown. 1993. "Insect conservation and extensive agriculture: The savanna of southern Africa." In K. J. Gaston, T. R. New, and M. J. Samways, eds., *Perspectives on Insect Conservation*. Intercept, Andover, England, pp. 75–95.

Speight, M. C. D. 1989. *Saproxylic Invertebrates and Their Conservation*. Nature and Environment Series, No. 42. Council of Europe, Strasbourg.

Thomas, M. C. 1995. "The Lacey Act and entomology outside the United States." *Oriental Insects* 29:429–431.

Turpin, T. F. 1992. *The Insect Appreciation Digest*. Entomological Foundation, West Lafayette, Ind.

U.S. Fish and Wildlife Service. 1991. *American Burying Beetle (*Nicrophorus americanus*) Recovery Plan*. Newton Corner, Mass.

U.S. Fish and Wildlife Service. 1992. *CITES. Appendices I, II, III to the Convention on International Trade in Endangered Species of Wild Fauna and Flora*. Department of the Interior, Washington, D.C.

U.S. Fish and Wildlife Service. 1994. *Endangered and Threatened Wildlife and Plants*. Department of the Interior, Washington, D.C.

Wilson, E. O. 1984. *Biophilia. The Human Bond with Other Species*. Harvard University Press, Cambridge, Mass.

Wilson, E. O. 1987. "The little things that run the world. The importance and conservation of invertebrates." *Conservation Biology* 1(4):344–346.

Wilson, E. O. 1993. "Biophilia and the conservation ethic." In S. R. Kellert and E. O. Wilson, eds., *The Biophilia Hypothesis*. Island Press, Washington D.C., pp. 31–41.

World Conservation Monitoring Centre. 1993. *1994 IUCN Red List of Threatened Animals*. B. Goombridge, ed. IUCN—World Conservation Union, Cambridge, England.

Yen, A. 1993. 13. "The role of museums and zoos in influencing public attitudes towards invertebrate conservation". In K. J. Gaston, T. R. New, and M. J. Samways, eds., *Perspectives on Insect Conservation*. Intercept, Andover, England, pp. 213–229.

Yen, A. L., and T. R. New. 1995. "Is invertebrate collecting a threatening process?" *Victorian Naturalist* 112(1):36–39.

ACKNOWLEDGMENTS

It is our hope that the readers of this book will be afforded a glimpse into a fascinating world seldom seen and appreciated. If you do not find this book appealing, or find it factually incorrect, we take full responsibility for the tone and errors found within. However, if after having read this book, you find yourself fascinated by beetles and are inspired to learn more about insects and their role in the natural world, then we must share the credit with the following individuals, for it is their generous assistance and expertise that made this book possible.

First and foremost, we would like to express our sincere appreciation to Peter N. Nevraumont of Nevraumont Publishing Company, who was instrumental in the conception and execution of this book. We are most grateful for his invaluable suggestions, patience, and steady encouragement throughout the project. Henry Galiano of Maxilla and Mandible, New York, first suggested the idea for the book and without him, the opportunity to write this book would not have existed. Henry and Luis Rivera, also of Maxilla and Mandible, produced and prepared many of the exquisite specimens photographed in the book. We thank Stepanie Hiebert for her skillful and thorough job of copyediting the manuscript. Her contribution added immeasurably to the readability of the book. The artistry of Lisa Charles Watson translated the natural beauty of preserved beetles into the stunning photographs seen here. Patricia Wynne's line drawings added the dimension of life to the beetles presented in the book. José Conde, of Studio Pepin, Tokyo, gave the book its overall visual impact. At Henry Holt Reference Books, Associate Publisher Kenneth R. Wright, provided early and enthusiastic support for this book.

The writings of Stephen R. Kellert and E.O. Wilson helped to bring into focus the benefit of beetles and the many levels at which they impact our lives. We are grateful to our fellow coleopterists, both living and dead, whose papers, correspondence, and discussions planted the seeds for the discussions presented in the book. Much of the information presented here is a result of their hard work and diligence. The authors are particularly grateful for the expert tutelage of three eminent coleopterists during their undergraduate and graduate days: Dr. Elbert Sleeper (retired) California State University, Long Beach, and Professors Clarke Scholtz and Erik Holm (retired), University of Pretoria, South Africa. It was through these men that we were afforded some of our first important glimpses of beetles on a truly global scale.

Colleagues who have been particularly generous with their knowledge include: William Barr, Josef Beierl, Svata Bílý, David Carlson, Cassandra Carter, Richard Cunningham, Robert Duff, Sebastian Endrödy-Younga, David Faulkner, Bruce Gill, Henry Hespenhiede, Frank Hovore, Henry and Anne Howden, Mary Liz Jameson, Paul Lago, Robert Moore, Randy Morgan, Brendan Moyle, Hans Mühle, Gayle Nelson, Paul Pearce-Kelly, Steve Prchal, Alex Reifschneider, Jacques Rifkind, David Russell, Kirk Smith, William Smith, Paul and Phyllis Spangler, William Warner, Richard Westcott, Dave Verity, the late George Vogt, Mark Volkovitch, George Walters, the late Don Whitehead.

Several of our friends and colleagues generously contributed specimens, photographs, artifacts, and anecdotes to the project. Dr. Rosser Garrison, Gerald Larsson, Paul McGray, Dr. Shephan Schaal of Naturmuseum Senckenberg, Dr. George Poinar and Dr. Brett Ratcliffe generously provided specimens and photographs for use in this work. Dr. Brian Brown and Brian Harris of the Entomology Section of the Natural History Museum of Los Angeles County provided us with many important and rare specimens, as well as cultural artifacts, many of which appear in this book. Dick Meier of the photography department at the Natural History Museum also provided photos. Peter Sloman contributed substantially to the culinary aspects of the book.

Another important layer of support is that which can only be supplied by friends and family. A special note of thanks is due to Joan and Sumner Ladd, who afforded to one of us their wonderful island paradise in Avalon, where the earlier drafts of this book were written.

We would like to express our deepest appreciation and thanks to our families. To our parents, Lois and Edwin Evans and Ivalou and Charles Bellamy, who have provided a lifetime of support. Most importantly, our wives, Mary Sullivan and Rose Bellamy, who have had to deal with our sometimes long absences and all too frequent beetle-driven distractions and yet still maintain a special understanding for our passion. Without them, even in a world filled with beetles, our lives would not be as rich.

Photographer's Acknowledgments

Special thanks to:
Stephan DeSantis for his many hours of hard work.
Navin Tiwari at CYMK Lab, NYC, for his generosity and professional services.
Peter N. Nevraumont for this photographic opportunity.
Wade Watson for his support and patience, and Max.

INDEX